Power Systems

Roberto Benato · Antonio Paolucci

EHV AC Undergrounding Electrical Power

Performance and Planning

Springer

Roberto Benato, Asst. Prof.
Università di Padova
Dipto. Ingegneria Elettrica
Via Gradenigo, 6/A
35131 Padova
Italy
benato@die.unipd.it

Antonio Paolucci, Prof.
Università di Padova
Dipto. Ingegneria Elettrica
Via Gradenigo, 6/A
35131 Padova
Italy

ISSN 1612-1287 e-ISSN 1860-4676
ISBN 978-1-4471-2553-2 ISBN 978-1-84882-867-4(eBook)
DOI 10.1007/978-1-84882-867-4
Springer London Dordrecht Heidelberg New York

British Library Cataloguing in Publication Data
A catalogue record for this book is available from the British Library

Cover design: deblik, Berlin, Germany
Typesetting and production: le-tex publishing services GmbH, Leipzig, Germany

Printed on acid-free paper

Springer is part of Springer Science+Business Media (www.springer.com)

to my three wonderful women Monica,
Isabel Maria and Letizia Maria

Roberto

to my wife, my son Marco and my daughter
Chiara

Antonio

Foreword

This is a time in which the deployment of HV and EHV insulated cable systems is exhibiting significant growth and, most importantly, when serious consideration is being given to the adoption of such systems in a much greater variety of cases than in the past.

What this implies is that insulated cables are currently looked upon by users as an enabling network solution that can help them address effectively, rapidly and reliably a variety of complex problems, for example through the deployment of mixed overhead and underground lines, rather than, as was often the case in the past, a niche solution to be employed only when absolutely inevitable.

This trend, driven by environmental concerns which have made overhead lines less attractive, and facilitated by technological progress which has made insulated cables more competitive, requires an ongoing objective evaluation of the possibilities offered by modern cable and mixed systems.

Against this background this book represents a timely and useful contribution towards a better understanding of the factors that need to be taken into consideration when introducing insulated cable systems in the network.

The book's thorough coverage of many of the associated technical aspects, coupled with its analysis of the broader economic, environmental and social factors involved, will certainly be of great interest for cable system users, cable engineers and students of electrical engineering.

Milan, May 2009

Aldo Bolza

Preface

We shall not cease from exploration
And the end of all our exploring
Will be to arrive where we started
And know the place for the first time.

T.S. Eliot, *Four Quartets*

In the development of the AC power systems, long transmission lines and overhead lines (OHL) have been considered for many years an indissoluble binomial, whereas the use of HV and EHV insulated cables was devoted mostly to DC submarine links; in the last decades, a strengthened sensibility towards the environment, with consequent big hindrances to OHL installations, and an increased reliability of high quality extruded insulations (cross-linked polyethylene – XLPE) of cables have induced the transmission system planners and grid owners to install numerous AC HV and EHV cable lines: they have unquestionable qualities of adaptability to the problems given by the territory when realizing modifications, widening or reinforcement of the transmission network. Nowadays, in the jargon of electrical engineering, words such as *undergrounding*, *UGC* (*U*nderground *C*ables) and *mixed lines* (a cascade composition of OHL and UGC) have a clear meaning.

The planning of any new line (OHL, UGC or mixed line) in the electrical grid must be always validated by extensive network simulations, among which power flow, short circuit, transient analyses; generally those studies ought to be preceded by a propedeutical investigation, in order to assist the system planner in the configuration, comparison and choice of the more suitable solutions, chiefly when the field of choice can be very wide as in the insulated cable market. In fact, a target of this book was the development of this propedeutical phase regarding the *electric energy transmission standpoint* with particular attention to the operating requirements.

Starting from the classical equations of transmission lines, developed more than a century ago, and by rearranging them with original procedures, the authors have created some novel capability charts (for UGCs in Chapter 3, for mixed lines in Chapter 4) which immediately visualize the best possible performances within the limits of current and voltage compatible with the line lifetime and power quality. These capability charts (which take into account the steady state ampacity and voltage levels, subtransient voltages, shunt compensation degree and other important parameters) are of primary importance (in the same way of power circle diagrams): they offer an original and efficient guide extremely useful for both the planning of

UGC or mixed line systems and a full compatibility of their insertion in the power network.

Through the book, there are also several occasions of cultural widening in the field of the circuit theory: one of the most meaningful contributions is the *recovery* of Ossanna's theory, completely developed in 1926 (and in the book newly presented in Section 3.9.2). It gives a simple but extremely elegant and powerful analytical direct computation of the possible voltage regimes obtainable in any AC power system by fixing the complex power at one of its ports, so avoiding the iterative procedures up till now largely used. This method, particularly useful to complete the line performance analysis, ought to be also generally appreciated as one of the more important tool in the circuit theory.

Also, the authors' attempt in Chapter 6 of framing the economical comparison between UGC and OHL on an engineering fair basis takes aim at helping the involved stakeholders to clear the field of subjective impressions.

Briefly, the book offers:

- a wide panorama of the large EHV cable installations which has been enhanced by means of some detailed reports of *paradigmatic* installations (Chapter 1);
- an outline of positive sequence modelling to analyse the steady state regimes of typical OHLs, of cross-bonded (with phase transpositions) cable systems and of gas insulated lines (Chapter 2);
- call to the constant use of some international rules and national standards;
- novel capability charts (with a lot of examples both for UGCs and mixed lines) which offer a throughout visualization of transmission line performances compatible with chosen current and voltage constraints;
- criteria regarding the energization and de-energization phenomena of no-load transmission lines (UGC and mixed lines);
- criteria for the dimensioning and check of the shunt reactive compensation of UGC (uniformly or lumped);
- comments of educational purpose on the physical meaning of some formal analytical expressions;
- introduction to the multiconductor analysis of undergrounding with self-made power frequency matrix procedures (Chapter 5), which allows investigating the electrical behaviours of all cable conductors (phases and sheaths) with some case studies and offers the possibility of comparing and validating simplified approaches;
- a technical and economical comparative procedure between overhead and cable lines which can contribute to the Environmental Impact Assessment recommended by the European community (Chapter 6).

The authors hope to have given the electric engineers a modern and sound tool to face the challenges of the future electrical grids where the undergrounding will play a key role.

Padova, June 2009 Roberto Benato and Antonio Paolucci

Contents

Chapter 1
HV Cable World Statistics and some Large Installations

1.1 Introduction

The National Academy of Engineering has rightly stated that the electrical grid is the greatest engineering achievement of the twentieth century [1].

The invisible presence of electricity and its grid in the industrialized societies is absolutely taken for granted in spite of its huge extension and possible vulnerability (blackouts!). When thinking of the invisibility of electricity, engineers turn their thoughts to cable undergrounding.

From the onset of the transmission and distribution of electrical power undergrounding has always played a key role.

It is suggestive to remember that the first electrical line supplying Pearl Street in New York (1882) constituted of DC underground insulated conductors laid in ducts under the street. Moreover, the first known public supply of electricity in Great Britain began in 1881 when a small river was harnessed to provide electrical power for lighting the streets of Godalming in the county of Surrey. At that time, there were no legal powers to enable would-be suppliers of electricity to bury cables under streets or footpaths. In consequence, the Godalming cables were laid in the street gutters [2].

Worldwide, the need for undergrounding installations became more and more pronounced with each passing year owing to an ever-growing concern for safety and amenities in densely inhabited urban and suburban areas as well as the preservation of aesthetic values in many localities.

The necessity of enforcing and making sturdier the electrical transmission network and the insurmountable difficulties in erecting new overhead lines force more and more towards the use in the power networks of undergrounding links.

At such level of maturity of the electrical grid the importance of cable systems is destined to strongly increase in the medium and long term.

Many issues [3] can be addressed to justify this asseveration:

- The ever-growing energy consumptions in conurbations of industrialized countries can be only meet by using underground cables;

- Great difficulties of erecting new overhead lines (even outside the large urban areas) due to a strong environmental awareness of the public;
- In some countries, the exposure limit of the magnetic fields will force the grid owners or operators to substitute parts of existing overhead lines with underground cables;
- Parts of the existing cable grid constituted of oil-filled paper insulation cables (installed up to the end of 1950s) must be replaced and re-investment in undergrounding will be needed in the next few years;
- In many developing countries (e.g. several regions of Turkey), there is a considerable pent-up demand in respect of reliable power supply systems;
- The Third World countries will surely expand their electrical grid by predominantly using cables.

In many of the above-mentioned situations, the electrical line is not more monotype but is a combination of overhead and cable line. This book will carefully deal with this new type of line: the mixed line.

Another issue deserves a very careful consideration.

The authors, when facing the question of land constraints for electric infrastructures, have often begun with this outstanding passage taken from an EPRI book [4] that postulates the necessity to develop common corridors in order to satisfy the ever-growing exchanges between areas of big industrialized systems:

"*Common Corridors*

For both ecological and economic reasons future expansion of transmission systems especially near urban areas will have to consider sharing their rights-of-way with other land users. Independent use of land by single utilities can lead to inefficient use of natural resources. Development of new rights-of-way in future years will have to consider the location of suitable highway networks, gas lines, sewage systems, nearby utility transmission lines, telephone lines, underground transmission systems, and recreational areas. Such development will require a coordinated effort by the government and industry. A current study in [5] outlines the approach that the electric utilities may follow to encourage full use of their transmission line corridors."

This EPRI idea regarding the USA, even if with wide territory resources, is strongly recommendable to many other countries.

This topic is so felt and argued in Europe that the International Council on Large Electric Systems (Cigré) has established a Working Group B1.08 entitled *Cable systems in multipurpose or shared structures* [6] and also a Joint Working Group B3/B1.09 *Application of long high capacity Gas Insulated Lines in structures* [7] to study the feasibility and the characteristics of these technical solutions.

The rationalization of fundamental services in the land, such as electricity and transport both via rail and via highway, represents a higher degree of consciousness in conciliating environmental issues and technology. Of course it is also a big technical challenge because it is not trivial that there is a full compatibility between possible different infrastructures. Moreover, there are a lot of projects of new galleries between neighbouring countries (separated by mountain chains) and these possibilities represent absolutely unique opportunities for integrating (railway or highway) transport and energy transmission within the same corridor or within the same structure [8–13].

This could strengthen the scarce transnational network between States whose electrical networks are responsible for giving high hindrances to an effective international electric market.

Sharing different nature structures and power transmission line within the same corridor represents an optimization of service rights-of-way: conciliating technological society needs respecting nature and territory seems to be the challenge of the twenty-first century.

Once again, multi-purpose structures can cause cable technologies to further take off.

1.2 Statistics of Cable Installed Lengths

Data on the lengths of AC underground cable and overhead line currently installed are important in order to have an idea of the consistency and the possible future developments of undergrounding in the power grid.

The following statistics are based on a Cigré Brochure [14] which has collected data by means of a questionnaire sent to the Study Committee members.

The statistics have been divided into five voltage ranges chosen in order to group together similar design and operational principles.

The chosen voltage ranges have been:

- 50–109 kV;
- 110–219 kV;
- 220–314 kV;
- 315–500 kV; and
- 501–764 kV.

The Brochure highlights that in some cases data capture proved difficult, particularly for countries with a multitude of small independent utilities (for example the USA and Germany).

In addition, the national systems for maintaining such data have been discontinued in some countries, since utilities have been released from state control and experienced reorganization, merger and acquisition.

However, the total length of underground cable circuit expressed as a percentage of the total circuit length is shown in Figure 1.1.

The results show that the large majority of circuits are overhead and undergrounding has still a great gap. The proportion of circuits that are underground ranges from 6.6% (50–109 kV) down to 0.5% (315–500 kV).[1]

[1] According with English literature, the decimal numbers are denoted with full stops.

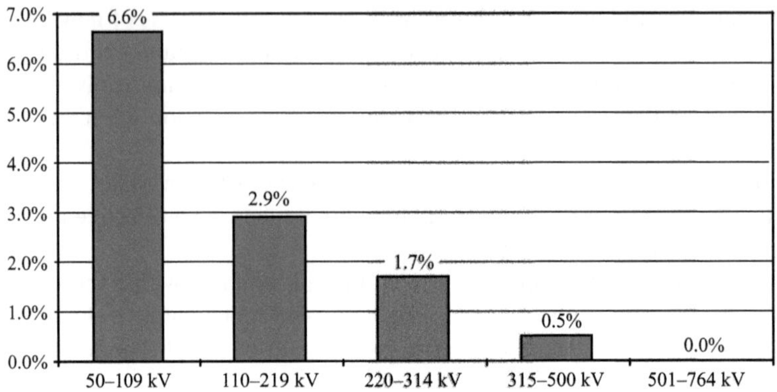

Figure 1.1 Percentage lengths of AC underground cable circuits on the total line length for each of the 5 voltage levels

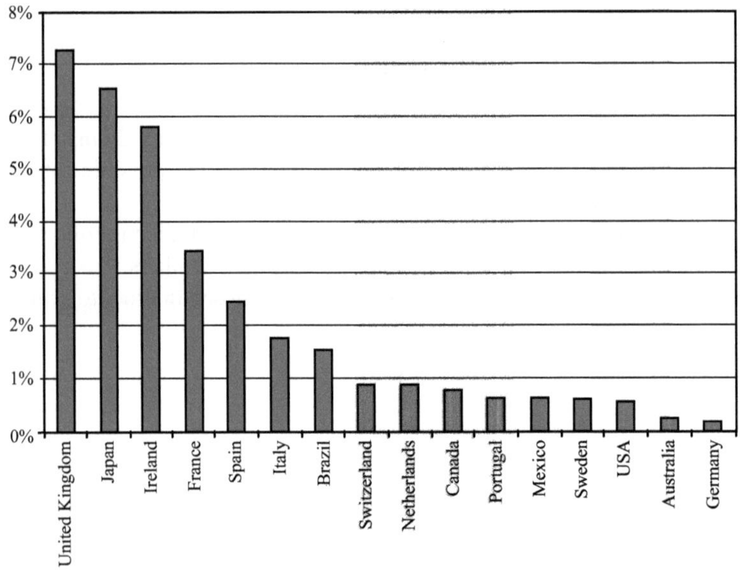

Figure 1.2 Percentage lengths of AC underground cable circuits on the total line length at HV (220–314 kV) for different countries

Figure 1.2 shows the percentage of underground circuit lengths installed in each country in the 220 to 314 kV voltage level whereas Figure 1.3 shows it in the 315 to 500 kV.

In order to simplify data collection, little technical detail was included in the questionnaire. However, where possible, respondents were asked to split the lengths of underground cable into those using lapped paper technologies and those using extruded polymeric insulation. The results are shown in Figure 1.4.

The decreasing proportion of extruded insulation used at the higher voltages reflects the incremental development of these cables.

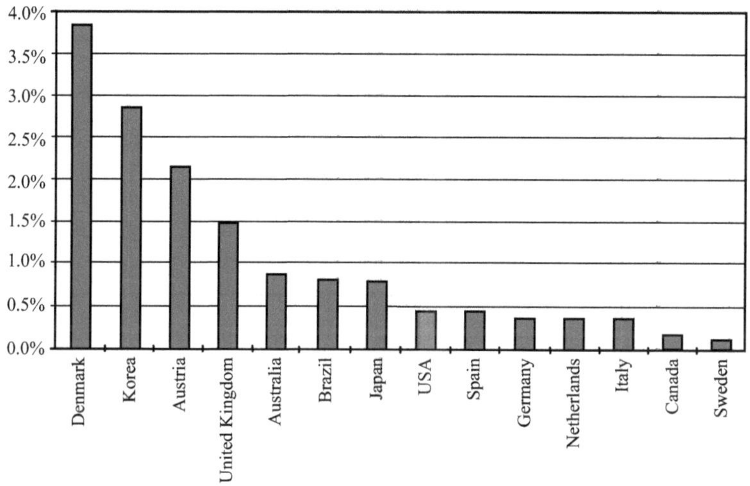

Figure 1.3 Percentage lengths of AC underground cable circuits on the total line length at EHV (315–500 kV) for different countries

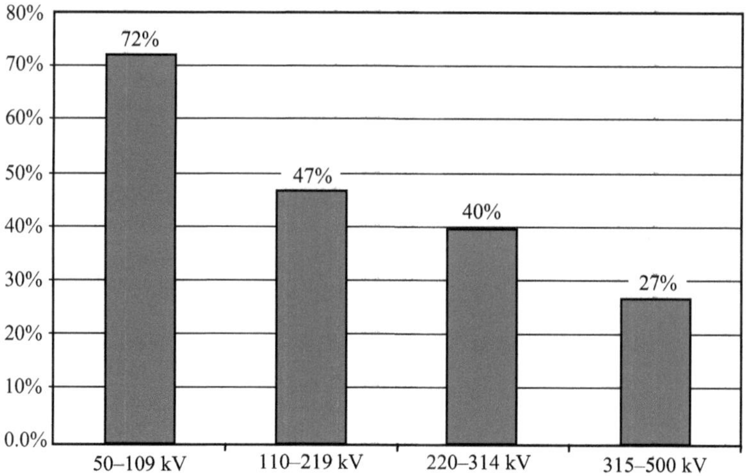

Figure 1.4 Percentage of the AC underground extruded polymeric-insulated cable

Lower voltage, low stress cables were developed first and, as the technology improved, extruded insulation was applied to higher voltage and used at higher stress.

Since the early 1960s 50 kV extruded cable has been in use, whereas 400 kV and 500 kV transmission circuits using extruded insulation were not introduced until around 2000.

The data on installed lengths clearly show that utilities have a strong preference for overhead lines rather than underground cables.

For the 50 kV to 109 kV range, 93% of the network is overhead.

In the voltage range 501–764 kV there are not cable installations.

Table 1.1 Major EHV Underground Projects (at power frequency) [15]

Country	Rated (phase-to-phase) voltage (kV)	Type and number of joints (1)	Type of installation (2)	Route length (km)	Number of circuits	Conductor cross-section (mm^2) Winter transmission total capacity (MVA)	Commissioning year
Denmark (Copenhagen Southern cable route)	400	CPFJ 72	DB	22	1	1600 Cu 975	1997
Denmark (Copenhagen Northern cable route)	400	PMJ 42	DB	12	1	1600 Cu 800	1999
Germany (Berlin/BEWAG Mitte-Friedrichshain)	400	CPFJ+PMJ 48	T	6.3	2	1600 Cu 1100	1998
Germany (Berlin/BEWAG Friedrichshain-Marzahn)	400	CPFJ+PMJ 30	T	5.5	2	1600 Cu 1100	2000
Japan (Tokyo (3))	500	EMJ 264	T	39.8	2	2500 Cu 2400 (4)	2000
United Arab Emirates (Abu Dhabi)	400	PMJ 12	D&M	1.3 (5)	4	800 Cu not available	2000
Spain (Madrid) Barajas	400	CPFJ+PMJ 96	T	12.8	2	2500 Cu 3440	2004
Denmark (Jutland)	400	PMJ 96	DB&D	14.5	2	1200 Al 1200	2004
United Kingdom (London)	400	CPFJ 60	T	20	1	2500 Cu 1600	2005
The Netherlands (Rotterdam)	400	PMJ 3	DB&D	2.25	1	1600 Cu 1000	2005
Austria (Wienstrom)	380	PMJ 30	DB&T&M	5.2	2	1200 Cu 1400	2005
Italy (Milan) (6)	380	PMJ 60	DB&D	8.4	2	2000 Cu 2100	2006

(1) CPFJ – Composite Prefabricated joint, PMJ – Premoulded Joint,
EMJ – Extruded Moulded Joint
(2) T – Tunnel, DB – Directly Buried, D – Ducts, D&M – Ducts and Manhole
(3) Cable system prequalified following Japanese Specification
(4) 1200 MVA/circuit with forced cooling in the future, 900 MVA/circuit now
(5) 15 core kms/4 circuits × 3 phases = 1.3 km
(6) See mixed line Turbigo–Rho in Section 1.6

Table 1.2 Design criteria adopted for 400 kV Projects [15]

Project	Cable circuit	Conductor	Electric stresses IN/OUT (kV/mm)	Metallic screen	Outer sheath (1)	Joints
Berlin	A	1600 mm^2 Cu 5 segments	11.5/5.4	Cu wires + Al laminated	PE with FRV	39 CPFJ
	B	1600 mm^2 Cu 5 segments	12.5/6.2	Cu wires + Al laminated	PE with FRV	15 PMJ
	C	1600 mm^2 Cu 6 segments	12.5/6.2	Corrugated Al	PE FRV	24 PMJ
Copenhagen		1600 mm^2 Cu keystone	11.5/4.9	Extruded Lead	PE with SCL	72 CPFJ 42 PMJ
Madrid (Barajas)	A	2500 mm^2 Cu 6 segments	11.6/6.5	Cu wires + Al laminated	PE with FRL	48 CPFJ
	B	2500 mm^2 Cu 6 segments	12.5/7.2	Al welded laminated	FR PE	48 PMJ
London		2500 mm^2 Cu 6 segments	11.6/6.5	Cu wires + Al laminated	PE with FRL	60 CPFJ
Jutland		1200 mm^2 Al stranded	12.6/6	Al wires + Al laminated	PE with SCL	96 PMJ
Rotterdam		1600 mm^2 Cu	11.8/5.9	Extruded lead	PE	6 PMJ
Vienna		1200 mm^2 Cu 5 segments	12.1/5.7	Cu wires + Al welded	PE	30 PMJ
Mixed line Turbigo–Rho		2000 mm^2 Cu 6 segments	11.8/6.4	Welded Al	PE	66 PMJ

(1) FRV – Flame retardant varnish; FRL – Flame retardant layer; SCL – semi conductive layer

1.3 Large Installations of EHV Cable Systems

In order to show the great development of EHV undergrounding in the last few decades it seems particularly meaningful to present the Tables 1.1 and 1.2 [15].

A lot of EHV cable installations are part of a longer overhead line (e.g. Turbigo–Rho, the Barajas project): once again the importance of mixed line arises.

1.4 Land and Submarine 150 kV AC Cable Link Sardinia–Corsica: SAR.CO

This installation can be considered as submarine HV AC reference installation [16].

Even if the islands of Sardinia and Corsica were partly connected through the 200 kV DC cable installed in the early sixties (called SA.CO.I. – Sardinia–Corsica–

Italy) [17], the need to improve power interchange between the two islands has required a new connection to satisfy the highly increased energy demands of these two tourist sites. The choice for the connection has been a 150 kV XLPE AC cable with 400 mm^2 copper conductor (as shown in Figure 1.6). The total length of the connection is 31 km and is composed of 10.5 km of land cable on Corsica, 15 km of submarine cable and 5.5 km on land in Sardinia (Figure 1.5). The nominal continuous rating is 150 MVA.

For the submarine part a three-core armoured cable has been selected (see Figures 1.6 and 1.7), while for the land part three single-core cables (Figure 1.7).

The maximum sea depth of the line is 75 m. The submarine cable (for constructional details see Figure 1.7) was buried in the sea bed for the major part of the route (about 70%) and most of all in the proximity of the seashore in order to prevent mechanical damage; additional protection has been applied in some particular cases.

The land cable was laid flat at 200 mm spacing and directly buried in trench at a depth of 1.4 m (see Figure 1.8). The metallic screens were cross-bonded.

The fibre-optics cable runs in a duct of the same trench: successively, it goes on embedded in the three-core submarine cable. It is worth noting that the design of the land and the submarine cable core is similar: this facilitates the transition joints between the single-core land cable and the three-core submarine one.

Before the installation of such a link, the 50 Hz AC Corsica grid was not synchronous with that of the Sardinia being linked to the SA.CO.I. system (200 kV DC) by means of the intermediate conversion station of Lucciana (North–East Corsica coast).

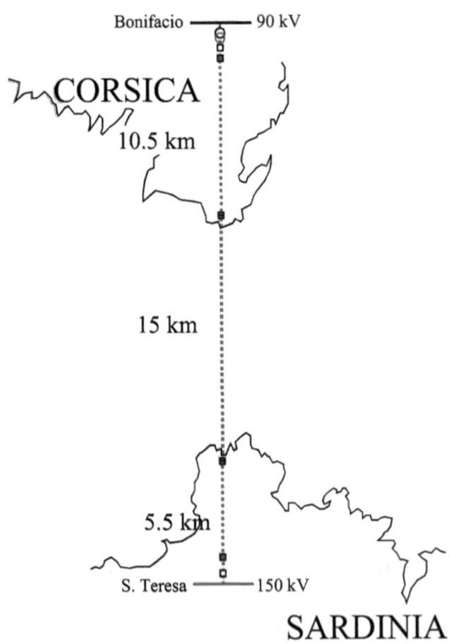

Figure 1.5 Sardinia–Corsica (SAR.CO) cable link

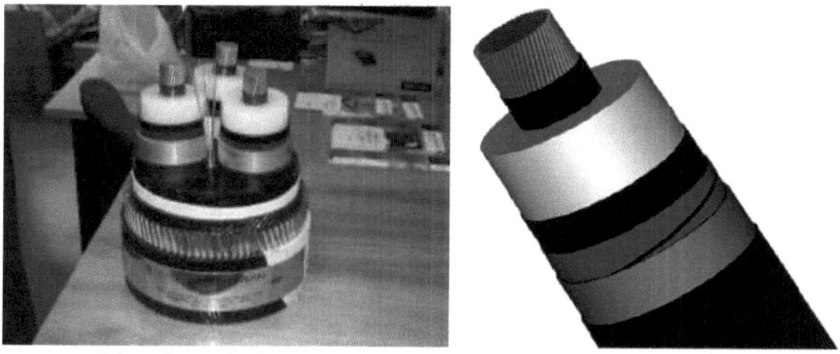

Submarine cable ϕ=207 mm Land cable ϕ=84 mm

Figure 1.6 AC three-core submarine extruded polymeric-insulated cable and single-core land one (SAR.CO) (Courtesy of Prysmian)

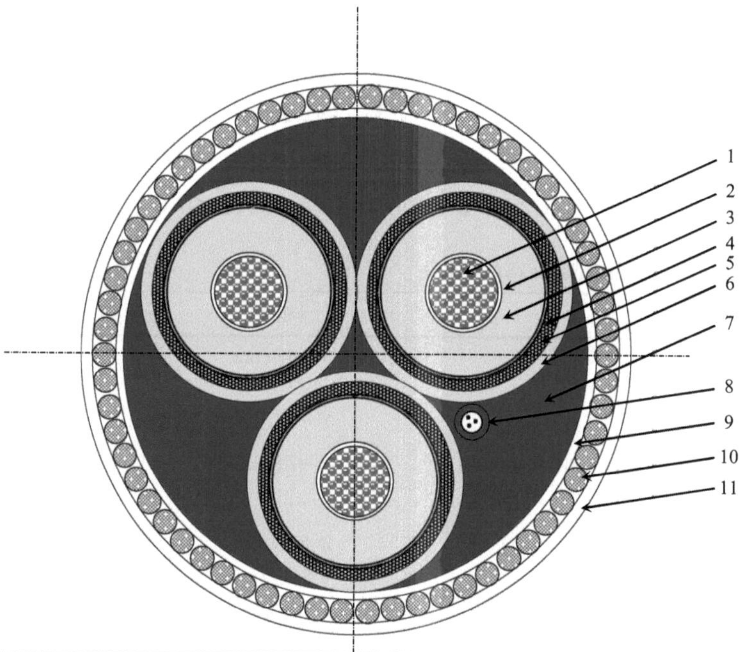

LEGEND

1) Copper conductor (400 mm^2)
2) Semi conductive screen
3) XLPE insulation
4) Insulation screen (semi conductive water swellable tapes)
5) Lead sheath
6) PE Sheath
7) Fillers
8) Fibre-optic cable
9) Reinforcement and bedding tape
10) Galvanized steel wires armour
11) Polypropylene yarn (ϕ=207 mm)

Figure 1.7 Constructive details of three-core submarine cable (SAR.CO)

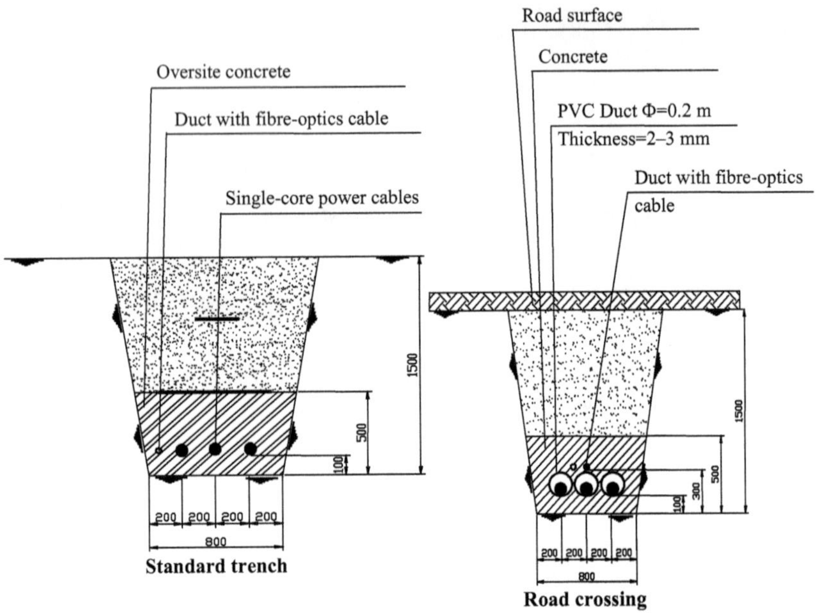

Figure 1.8 Installation details of land cable ($\phi = 83.9$ mm) laying (SAR.CO) (Courtesy of Terna)

Table 1.3 Power cable characteristics (SAR.CO)

	Land	Submarine
Type	Single-core	Three-core
Cross-section (copper)	400 mm^2	400 mm^2
Copper compacted watertight	ϕ 25.1 mm	
Semi conductive layer		
Cross-linked polyethylene (XLPE) insulation	ϕ 67.8 mm	
Water barrier (hygroscopic tapes)		
Al tape longitudinally welded		
PE with graphite coating	ϕ 83.9 mm	ϕ 207 mm
Cable mass	8.6 [kg/m]	75 [kg/m]
Maximum voltage for the equipment	170 kV	170 kV
Operating phase-to-earth voltage	87 kV	87 kV
Operating phase-to-phase voltage	150 kV	150 kV
Insulation voltage level (lightning)	750 kV	750 kV
Short-circuit withstand current	20 kA/0.5 s	20 kA/0.5 s
Route length	5.5 + 10.5 km	15 km
Ampacity	580 A	580 A
Transmission capacity at 150 kV	150 MVA	150 MVA
Phase conductor DC resistance at 20 °C	0.047 Ω/km	0.047 Ω/km
Sheath DC resistance at 20 °C	0.15 Ω/km	0.35 Ω/km
Shunt capacitance	0.18 μF/km	0.15 μF/km
Loss factor at maximum operating temperature	0.001	0.001

Table 1.3 collects a lot of data about the design of two cable types used in the SAR.CO installation.

The submarine cable route has taken into account the existing 200 kV DC SA.CO.I cable route: starting from San Teresa di Gallura in Sardinia and arriving in Corsica at Cala Sciumara, the 150 kV AC cable crosses the SA.CO.I 200 kV DC cables only one time and in the remaining part of the route is laid at a distance of about 400 m.

1.5 The Madrid "Barajas" Airport Project (Spain)

This project can be considered as a reference tunnel installation [18, 19].

AENA, the Spanish airport authority, had to extend the international Barajas Airport of Madrid.

This extension included the construction of two new runways.

Since the existing 400 kV overhead transmission lines operated by REE would have obstructed the incoming airport planes, they had to be dismantled and replaced by two cable (each with 2500 mm^2 Copper conductor) circuits installed in a tunnel crossing under the new runways.

Figure 1.9 shows (inside the circle) the double-circuit cable link (12.8 km) in the electrical map of a part of the Spanish grid.

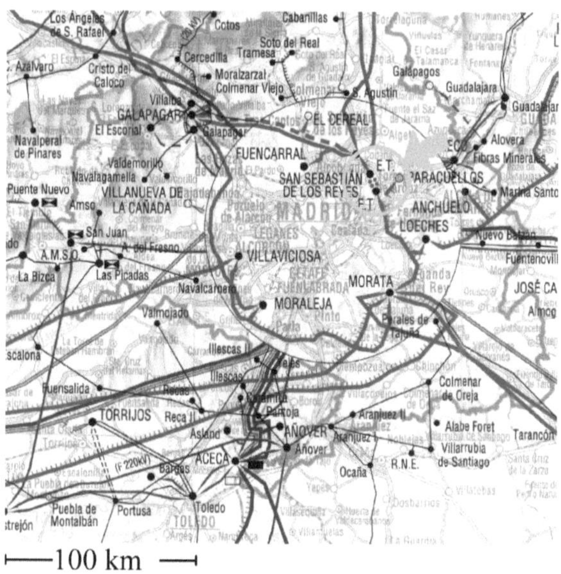

Figure 1.9 Electrical map with the indication of the double-circuit cable (Barajas: 12.8 km) (Courtesy of RED Eléctrica de ESPAÑA)

1.5.1 Milestones of the Barajas Project

2001	*Economical agreement between REE and AENA*
June 2002	*Presentation of the project for authorization procedure*
2003	*Beginning of the cable laying with the tunnel still un-concluded (cable laying concluded in January 2004)*;
February 2004	*In service*

In order to respect the foreseen timetable, the yard could count on up to 200 skilled worker group simultaneously, with five Pirelli jointer teams and four ABB ones present together on site.

1.5.2 Technical Characteristics of the Link

In a part of the route the two single-circuit overhead lines, San Sebastian de los Reyes to Moorata (35 km long) and San Sebastian de los Reyes to Loeches (25 km long), became a double-circuit. In this part, the overhead lines have been substituted by a double-circuit cable i.e. by an underground link 12.8 km long.

The six single-core cables of the double-circuit are installed in the same tunnel with the possibility of forced air cooling. The double-circuit is not electrically in parallel.

Each cable circuit contains 48 pre-fabricated joints, sectionalized, straight and earthed, and six outdoor terminations plus all required bonding cable and Surge Voltage Limiters (SVLs) for the cross bonding arrangement. The ABB joints are equipped with capacitive sensors in order to perform partial discharge measurements.

Additional current transformers and surge arresters for differential protection of the total cable system are also installed.

Table 1.4 sums up the chief characteristics of the installation.

The bonding of sheaths consists (see Figure 1.10) of a combination of 5×3 minor cross-bonding sections (810 m long) and two end single-point bonding sections of 300 m and 400 m long. These two sections have allowed to absorb the possible route variations (with respect to the project) and to order 90 drums of equal length.

The cross-bonding is performed without phase transpositions.

Even if the winter ampacity reported in Table 1.4 is rather high (i.e. 2482 A) the link is at present operated at an average current of about 600 A per circuit.

1.5.3 Tunnel and Earthing System Characteristics

The tunnel is buried at about 2 m depth and in a part of the route runs parallel to a river so that some pumps have been installed (25 l/s) for the possible water extraction in emergency conditions.

Table 1.4 Barajas installation characteristics

Maximum voltage for the equipment	U_m	420 kV
Phase-to-phase operating voltage	U_N	400 kV
Phase-to-earth operating voltage	U_0	230 kV
Withstand short circuit		50 kA/0.5 s
Voltage insulation levels	Waveform 250/2500 μs	1050 kV
	Waveform 1.2/50	1425 kV
Route length		12.8 km
Ampacity	(Winter)	2 × 2482 A
Transmission capacity at 400 kV	(Winter)	2 × 1720 MVA
Joule power losses in the double-circuit	(Winter)	448.6 W/m
Ampacity	(Summer)	2 × 2006 A
Transmission capacity at 400 kV	(Summer)	2 × 1390 MVA
Joule power losses in the double-circuit	(Summer)	294.7 W/m
Inlet air max. temperature	(Summer)	+42 °C
	(Winter)	+25 °C
Max temperature of the air in the tunnel		+50 °C
Maximum forced cooling air speed		5 m/s

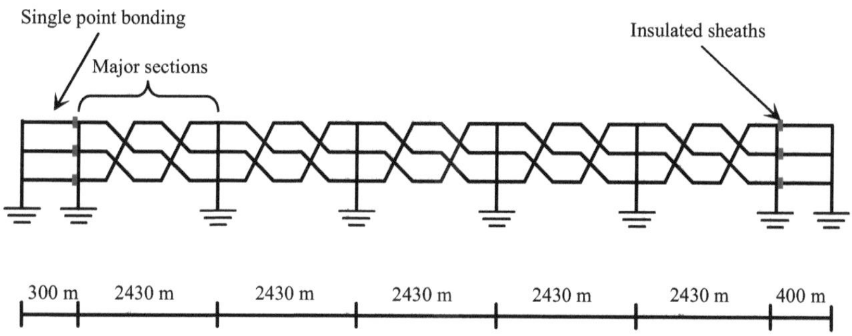

Figure 1.10 Sketch of the sheath bonding and earthing (Barajas)

In a location, the route has required a crossing of the river by means of a circular section viewable tunnel. The overall digging volume has been equal to 650 000 m^3. The tunnel has been constructed by 6500 pre-cast reinforced concrete units with inner dimensions 2 × 2.25 m (length × height) and thickness of 250 mm.

Some photographs during tunnel construction are shown in Figure 1.11.

For the earthing grid, the steel reinforcement of the gallery has not been used; it has been preferred a formation of four copper conductors ($\phi = 120\,\text{mm}^2$ and $\phi = 35\,\text{mm}^2$) located at the tunnel vertexes; a metallic ring bonds the four earthing conductors at regular intervals (400 m).

The tunnel has several technological equipments e.g. closed-loop camera, smoke detector system and telephone repeaters (see Figure 1.12) for maintenance and inspections.

Figure 1.11 Tunnel under construction (Barajas) (Courtesy of RED Eléctrica de ESPAÑA)

1.5.4 Power EHV Cables

The Barajas power cables have a cross section of 2500 mm^2 and their characteristics are reported in Table 1.5. The two manufacturers ABB and Pirelli have delivered a circuit each. Some photographs of Barajas samples are shown in Figure 1.13.

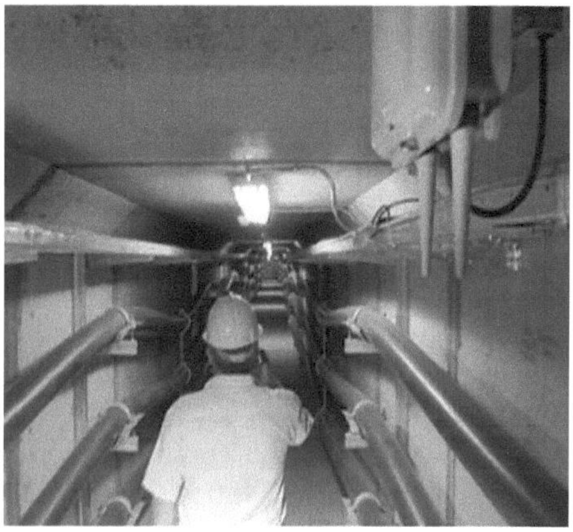

Figure 1.12 Detail of the telephone system (Barajas) (Courtesy of RED Eléctrica de ESPAÑA)

Table 1.5 Power cable characteristics (Barajas)

Manufactures	ABB	PIRELLI
Cross-section of single-core Copper cable	2500 mm^2	2500 mm^2
Milliken 6 sectors conductor	ϕ 65 mm	ϕ 65 mm
Semi conductive layer	ϕ 70 mm	ϕ 71.6 mm
Cross-linked polyethylene (XLPE) insulation	ϕ 125.9 mm	ϕ 122 mm
Semi conductive layer	ϕ 129.7 mm	ϕ 126 mm
Water swelling tape	–	ϕ 128.3 mm
Aluminium sheath	–	ϕ 138.8 mm
Copper wires + aluminium foil	ϕ 130.9 mm	–
PE outer jacket	ϕ 148 mm	ϕ 142.5 mm
Cable mass	37 [kg/m]	40 [kg/m]
Drum length	810 m	
Numbers of drums	90	
Drum mass	$\approx 35 \times 10^3$ kg	
Drum dimensions	$\varnothing = 4.4$ m; width = 3 m	

Figure 1.13 Two samples of the "Barajas" Pirelli EHV cable (Courtesy of Prysmian)

1.5.5 Cable Laying in the Tunnel

Figure 1.14 shows a suggestive sight of the double-circuit in tunnel whereas Figure 1.15 offers a sketch of the same sight with the detailed indications of the spacings and tunnel dimensions: the cable spacing is 0.5 m whilst the circuit spacing is 1.44 m.

The cables are fixed at the tunnel lateral walls every 6 m with a snaking of 0.25 m: cables rest on saddles and brackets.

Due to the mechanical withstand to the short circuit electrodynamic forces, every 3 m the cable are tied together by means of three phase spacers (see Figure 1.16).

The four inlet forced cooling stations (a Pirelli project) are located along the tunnel at a distance of 2480 m (see Figure 1.17): each station is equipped with three

Figure 1.14 Double-circuit cable in tunnel (Barajas) (Courtesy of RED Eléctrica de ESPAÑA)

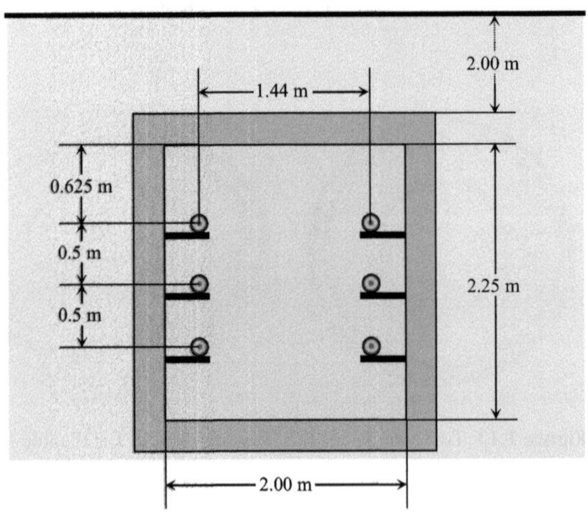

Figure 1.15 Sketch of the double-circuit cable in tunnel with indication of the spacings (Barajas)

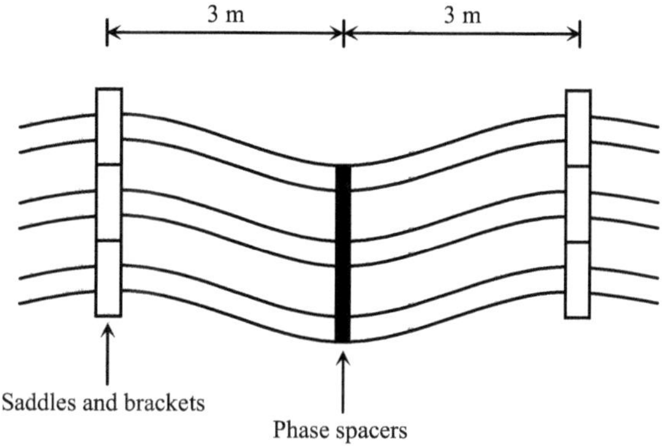

Figure 1.16 Cable snaking (Barajas)

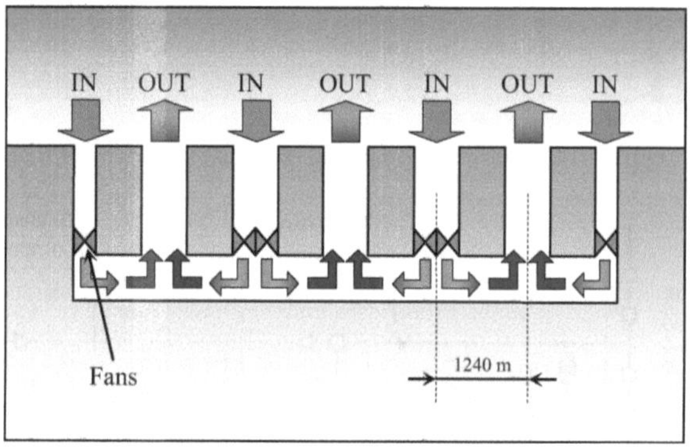

Figure 1.17 Forced cooling system (Barajas)

fans (38.3 kW rated) with a maximum air speed of 5 m/s. The outlet cooling stations are three.

The temperature control is driven by the RTTR (Real-Time Thermal Rating) system.

1.5.6 The Transition Compounds and Protection Schemes

The two transition compounds cover an area of $60 \times 40\,\mathrm{m}^2$ and are not equipped with disconnectors or circuit breakers hosting only the oil outdoor terminations and the surge arresters (see Figure 1.18). The outdoor terminals are supplied by the two manufacturers, whereas the six surge arresters are from ABB. The current transformers for the differential relays are toroidal-type.

The transition compound with a unique earthing grid (designed for 50 kA short circuit current) hosts a civic building ($8 \times 2.5 \times 2.75\,\mathrm{m}^3$) and an emergency diesel generator 250 kVA rated (other five emergency diesel generators have been installed in the five sites of forced air adduction).

Figure 1.18 Transition compound (Barajas) (Courtesy of RED Eléctrica de ESPAÑA)

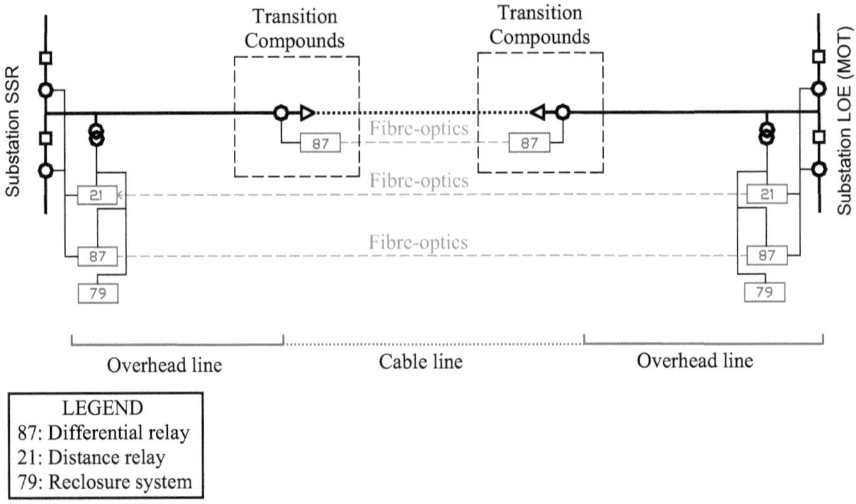

Figure 1.19 Relay architecture (shown only for one circuit) (Barajas)

In the utility building there are a 25 kV/400 V transformer, service switchboard, telecommunication and cable protection panels, tunnel monitoring terminals (closed-loop cameras, fibre-optics terminal, Data Thermal System and Real Time Thermal Rating).

The protection architecture is shown in Figure 1.19.

With regard to spare units, in one area given by the Airport authorities three drums and five joints for each circuit have been stored. In the ventilation station, a spare fan is also present.

Successively to the conclusion of the project, in the electrical stations of San Sebastian and Morata it has been necessary to install two three-phase shunt reactors of 150 Mvar in order to perform a good operation of the link.

1.6 380 kV Double-Circuit Cable of Mixed Line Turbigo–Rho (Italy)

This underground line can be considered as a directly buried reference 400 kV installation [20, 21].

The need of a 380 kV transmission line between the important network-node of Turbigo (reinforced by a 800 MW power plant) and the substations of Ospiate and Bovisio (which play a fundamental role in the power delivery of Milan City network) had been felt from several years, aiming at avoiding congestions subsequent to the high power flow in that area and reinforcing the West-East transmission axis in the Northern Italy.

The Turbigo–Rho link can be considered as a mixed line composed of a 20 km overhead line from Turbigo to Pogliano Milanese and 8.4 km double-circuit cable line from Pogliano Milanese to Rho: the undergrounding was needed in order to cross both strongly populated areas and natural reserves.

In order to fulfil the thermal power rating of the overhead line, it was necessary to install an underground double-circuit cable with a 2000 mm^2 copper conductor.

The two circuits were directly buried in two separate trenches (circuit spacing equal to 6 m) at a depth of 1.5 m, flat-type with 0.35 m cable spacing: the 6 m circuit spacing allows giving a good independence of the two circuits both in the thermal regimes and for maintenance purpose.

Outdoor terminations were installed in the overhead/underground transition compound. Continuous temperature monitoring with a DTS system and continuous partial discharge monitoring at joints were also installed.

A special magnetic field shielding has been provided for some parts of the route in accordance with the Italian laws [22, 23].

Figure 1.20 shows the electrical map with the indication of the mixed line route.

The project construction started in July 2005 and has been in service since 30 June 2006.

1.6.1 Milestones of the Turbigo–Rho Project

September 2004	*Authorization to the construction*
July 2005	*Opening of the yard*
30 June 2006	*In service*

It is worth noting that the short time for the realization (only 12 months) has been driven by a very precise schedule of the digging times.

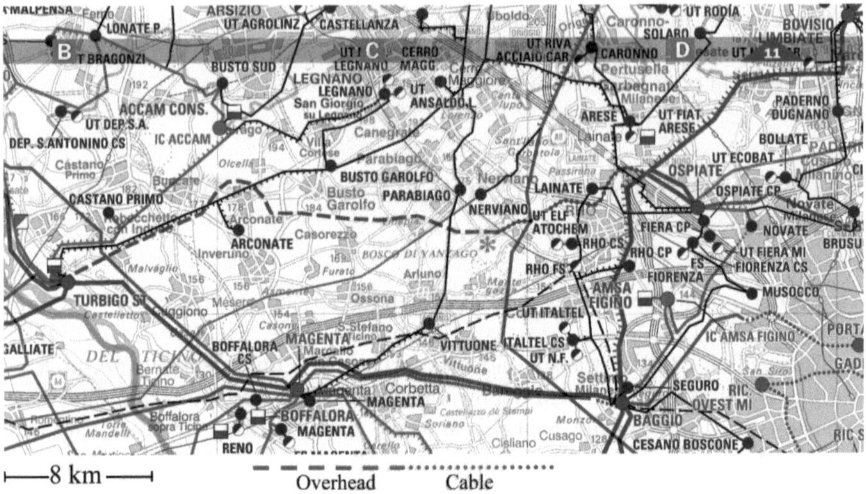

Figure 1.20 Electrical map of Turbigo–Rho 380 kV mixed line

Table 1.6 Turbigo–Rho installation characteristics

Maximum voltage for the equipment	U_m	420 kV
Phase-to-phase operating voltage	U_N	380 kV
Phase-to-earth operating voltage	U_0	220 kV
Withstand short circuit		50 kA/0.5 s
Voltage insulation levels	Waveform 250/2500 μs	1050 kV
	Waveform 1.2/50	1425 kV
Route length		8.4 km
Ampacity		2 × 1600 A
Transmission capacity at 380 kV		2 × 1053 MVA
Joule power losses in the double-circuit		245 W/m
Laying depth (to bottom of the trench)		1.5 m
Laying arrangement		Flat
Phase spacing		0.35 m
Circuit spacing		6 m

1.6.2 The Undergrounding Link of the Turbigo–Rho Mixed Line

For the underground part, the choice of the cable cross-section has been driven by the necessity of avoiding, in any possible operating condition, a bottleneck from a power rating standpoint. Since the overhead line can carry at thermal limit (in the cold months) a current of 3200 A (in accordance with the Italian Standard CEI 11-60 [24]), a double-circuit cable with 2000 mm^2 cross-section has been chosen.

Table 1.6 reports all the chief installation characteristics at a glance.

Another important feature of the installation regards the partial discharges (PD) monitoring. All installed accessories have an embedded capacitive sensor and are permanently connected through a special cable to the monitoring system. Today there are no standards for PD measurement on-site, mainly due to external and background noise level that can be neither foreseen nor limited below a given threshold. All accessories are connected to several local acquisition units, and all the local units are connected to an Ethernet LAN in S-ring configuration. Such configuration allows continuity in PD monitoring activity even if one or more channels of one or more acquisition units should experience any problem. In order to continuously monitor the performances and the behaviour of such an important connection, the system is provided with a variety of monitoring systems, e.g. a DTS system for thermal monitoring which collects all the information about the cables physical variables by means of suitable sensors (in this case, fibre-optic cables laid on top of the hottest phase). However, the most innovative of them is the above-mentioned permanent monitoring of partial discharges.

1.6.3 Power EHV Cables

The solution chosen for the underground part is a single-core *XLPE*-insulated cable designed for the voltage levels of Table 1.6.

The cable (see Figure 1.21) has a Milliken type (with six segments) copper conductor, with a cross section of 2000 mm^2, longitudinally watertight by means of hygroscopic tapes.

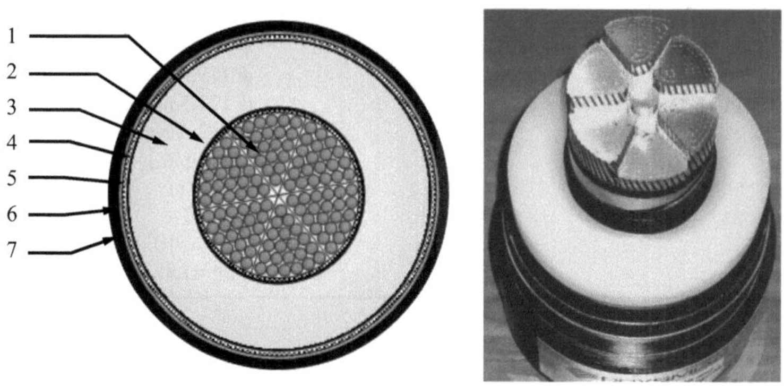

LEGEND
1) Cooper Milliken conductor (water-blocked)
2) Semi conductive screen
3) XLPE insulation
4) Semi conductive screen
5) Semi conductive water swellable tapes
6) Welded Aluminum Sheath
7) PE outer sheath

Figure 1.21 A sketch and a sample of the underground part of the "Turbigo–Rho" EHV mixed line (Courtesy of Prysmian)

Insulation consists of extruded cross-linked polyethylene (XLPE) suitable for operation at conductor temperature equal to 90 °C.

Under the metallic sheathing, a longitudinal water barrier is applied in order to limit the water penetration along the power core in case of cable damage.

The cable is provided with a longitudinally welded aluminium sheath, thick enough to withstand the rated phase-to-ground short circuit current (i.e. 50 kA per 0.5 s).

The metallic sheath acts also as a radial water barrier. The anti-corrosion protection consists of an extruded PE sheath. The cable outer sheath has an overall thin layer of graphite to permit the after laying voltage test of the sheath.

Table 1.7 reports other characteristics.

Metallic sheaths of the cables are electrically connected in a cross bonding system. The cross-bonding is performed with phase transpositions (see Figure 1.22).

Each phase consists of 12 cable spans with four major cross bonding sections composed of 710 m long minor sections.

The minor sections are connected using sectionalized joints. At these points the cable sheaths are cross-bonded via suitable link boxes, provided with SVLs.

Table 1.7 Power cable characteristics (Turbigo–Rho mixed line)

Manufacture	PRYSMIAN
Cross-section of the single core copper cable	2000 mm^2
Milliken 6 segments conductor	ϕ 57 mm
Semi conductive layer	ϕ 61.2 mm
Cross-linked polyethylene (XLPE) insulation	ϕ 113.8 mm
Water swelling tape	ϕ 119.3 mm
Aluminium sheath	ϕ 121.3 mm
PE outer jacket	ϕ 132.3 mm
Cable mass	31 [kg/m]
Drum length	710 m
Numbers of drums	72
Drum mass	$\approx 23.7 \times 10^3$ kg
Drum dimensions	$\varnothing = 4.2$ m; width = 2.5 m

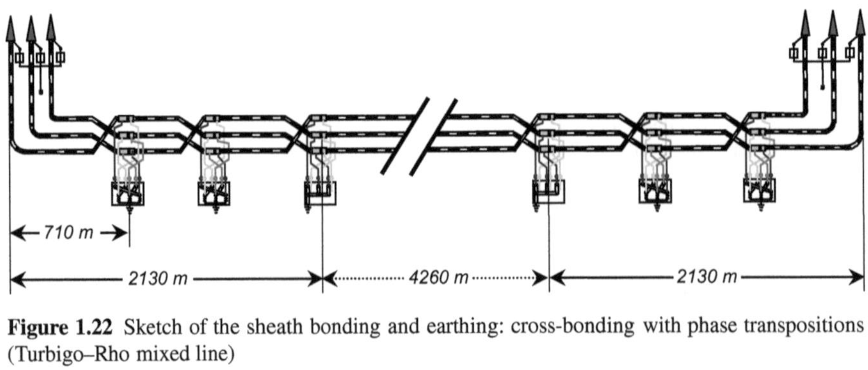

Figure 1.22 Sketch of the sheath bonding and earthing: cross-bonding with phase transpositions (Turbigo–Rho mixed line)

Straight joints are used to connect the major sections. At these points the cable sheaths are solidly bonded and earthed through three-core disconnecting link boxes.

At the terminations, the cable sheaths are connected to earth using disconnecting single-core link boxes.

Joints are of premoulded type, with the electrical part consisting of a single piece sleeve. As already mentioned, all accessories have been provided with a capacitive sensor in order to allow partial discharge monitoring.

1.7 Cable Laying

The cables are chiefly running parallel to roads of local authorities property (Province of Milan, Municipalities) with many crossings (roads, rivers, gas pipes, medium voltage cables, railways, etc.).

As already mentioned, the cables are installed in two separate trenches located on the opposite sides of the roads, 6 m spaced out.

With regard to the main road installation configuration, the cables are laid in flat formation in a sand backfill at 1.6 m depth protected by concrete slabs as in Figure 1.23.

With regard to the road crossings, the "open trench" technique has been adopted: the trench is similar to that of Figure 1.23 but the cables are laid inside polyvinyl chloride (PVC) pipes blocked in flat formation by concrete.

For river and road crossings the directional drilling technique has been used: the cables are laid inside PE pipes. For the crossing of rivers Olona and Lura the pipes

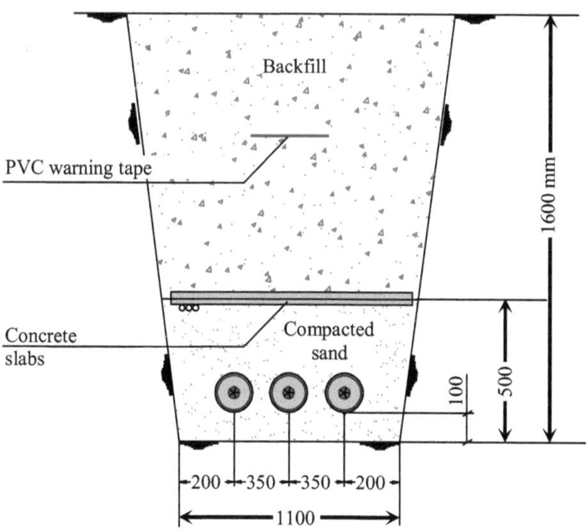

Figure 1.23 Standard cable laying in trench (Turbigo–Rho mixed line)

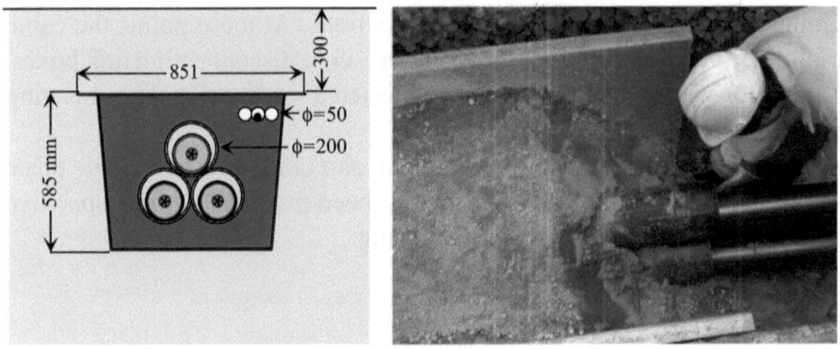

Figure 1.24 (Turbigo–Rho mixed line) Cable laying for the reduction of magnetic field (screening raceway) (Courtesy of TERNA)

Figure 1.25 (Turbigo–Rho mixed line) Cable drum (Courtesy of TERNA)

Figure 1.26 (Turbigo–Rho mixed line) Pulling winch and cable laying (Courtesy of TERNA)

Figure 1.27 (Turbigo–Rho mixed line) Joint location and trench (concrete slabs are also visible) during installation phase (Courtesy of TERNA)

are filled with water: in order to improve the heat dispersion, these pipes are suitably fitted with expansion tanks.

Another important constraint that the installation has overcome is the limit level of 3 μT imposed to Electromagnetic Field (EMF) by the Italian law [23] for new transmission lines at sensible receivers (i.e. where human presence lasts more than four hours per day). A special screening solution has been adopted in order to minimize the magnetic field in the sensitive areas: the cables are installed in a trefoil configuration inside single plastic ducts which are embedded in a special high magnetic permeability steel raceway suitably protected against corrosion (see Figure 1.24). This ferromagnetic raceway is composed of different elements: the hull, the cover and the closing clips. The design of the described open raceway combines the closed perimeter shielding efficiency with an open shape and the absence of welding. The particular shape allows also the raceways to follow the curves of the trench and lateral and vertical variations of direction. It is worth underlining that the dimensions of the raceway are tailored for this project. Indeed they have to reach the optimum compromise between the magnetic field screening effect and the minimization of the derating factor for the power cables.

Figures 1.25–1.27 show different phases during installation.

1.7.1 The Transition Compounds and Protection Schemes

Figure 1.28 shows the transition compound (which realizes the scheme of Figure 1.29) where there are not circuit-breakers but only disconnecting devices. In the same Figure 1.28 it is possible to see the surge arresters.

Figure 1.28 (Turbigo–Rho mixed line) Transition compound (Courtesy of TERNA)

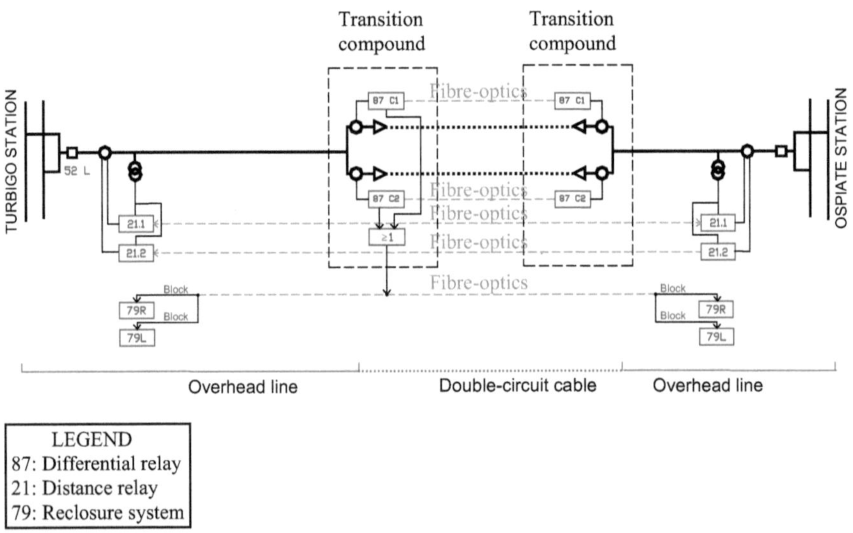

Figure 1.29 (Turbigo–Rho mixed line) Relay architecture

Another requirement has been to assure the highest level of reliability, also considering the protection systems currently in use for 380 kV. In particular the selectivity of the protection devices for rapid and slow re-closure cycles had to be maintained: in case of permanent faults (on the cable section) the line shall not be re-closed, but for faults on the overhead section (generally transient events) the cycle has to be completed.

All these features are present in the relay architecture scheme of Figure 1.29.

The different relays can detect whether or not the fault is on the cable section. If so, they block the (slow and fast) re-closure cycles.

References

1. P.F. Schewe: The grid: A journey through the heart of our electrified world, 2007, The National Academies Press
2. J.D. Endacott: Underground power cables, *Philosophical Transactions of the Royal Society of London*. Series A, Mathematical and Physical Sciences, Vol. 275, No. 1248, pp. 193–203, 1973
3. E. Peschke, E. von Olshausen: Cable systems for high and extra-high voltage, 1999, Publicis MDC Verlag, Germany
4. Electric Research Council – Electric Power Research Institute: Transmission Line Reference Book 345 kV and above, Palo Alto, California, 1975
5. Shared Transmission Corridors Aid Grid Growth, Electric Light and Power, March 1972
6. Cigré Technical Brochure No. 403, Working Group B1.28: Cable systems in multipurpose or shared structures, February 2010, ISBN: 978-2-85873-090-2
7. Cigré Technical Brochure No. 351, Working Group B3/B1.09: Application of long high capacity gas-insulated lines in structures, October 2008, ISBN: 978-2-85873-044-5
8. R. Benato, L. Fellin, A. Paolucci: Installation of GIL in railway tunnels, *Ingegneria Ferroviaria*, Year LVII, No. 7, luglio 2002, pp. 551–557 (in Italian)
9. R. Benato, P. Brunello, E.M. Carlini, C. Di Mario, L. Fellin, G. Knollseisen, M. Laußegger, M. Muhr, A. Paolucci, W. Stroppa, H. Worle, R. Woschitz: Italy–Austria GIL in the new planned railway galleries Fortezza–Innsbruck under Brenner Pass, *Cigré 2006, Paper B1-304* also in *E&I Elektrotechnik und Informationstechnik*, Vol. 123, No. 12, December 2006, pp. 551–558
10. R. Benato, C. Di Mario, H. Koch: High capability applications of Long Gas Insulated Lines in Structures, *IEEE Trans. on Power Delivery*, Vol. 22, Issue 1, pp. 619–626, January 2007
11. R. Benato, E. M. Carlini, C. Di Mario, L. Fellin, A. Paolucci, R. Turri: Gas Insulated Transmission Lines in Railway Galleries, *IEEE Trans. on Power Delivery*, Vol. 20, Issue 2, April 2005, pp. 704–709
12. R. Benato, E. M. Carlini, C. Di Mario, L. Fellin, G. Knollseisen, M. Laußegger, M. Muhr, H. Wörle, R. Woschitz: Gas Insulated Transmission Lines in Railway Galleries – Part II, *Proceedings of IEEE St. Petersburg Power Tech'05 Conference*, 27–30 June 2005, San Petersburg, Russia
13. R. Benato, M. Bernocchi, L. Camilli, C. Di Mario, C. Fourment, G. Roinel, F. Lesur: Compatibility between electrical energy transmission lines and new railway infrastructures, *Cigré 2008, Paper B1-308*
14. Cigré Technical Brochure No. 338, Working Group B1.07: Statistics of AC underground cables in power networks, December 2007
15. P. Argaut, K. Bjorlow-Larsen, E. Zaccone, A. Gustafsson F. Schell, V. Waschk: Large projects of EHV underground cable systems, *Paper A.2.1, JICABLE 2007* also in *Revue de l'électricite et de l'électronique*, No. 6/7, 2007, pp. 65–75

16. C. Di Mario, A. Gariel, M.R. Guarniere, R. Pintor, M. Rebolini, R. Rendina, P. Zanuzzi: New submarine cable link between Sardinia and Corsica (SAR.CO), *L'Energia Elettrica*, Vol. 83, No. 1/2, 2006, pp. 37–44 (in Italian)
17. P. Gazzana Priaroggia, G. Palandri, 200 kV D.C. Submarine Cable Interconnection between Sardinia, Corsica and Italy, *Cigré 1968, Paper 21-05*
18. R. Granadino, J. Planas, M. Portillo: Undergrounding the first 400 kV transmission line in Spain using 2500 mm^2 XLPE cables in a ventilated tunnel: the Madrid "Barajas" airport project, *JICABLE '03, Paper A.1.2*, 2003
19. P. Corsaro, R. Gaspari: The Barajas Airport Project: How a 400 kV Cable System Can Be Safely And Fully Exploited, *2004 International Conference on Power System Technology – POWERCON 2004*, Singapore, 21–24 November 2004
20. R. Rendina, A. Posati, M. Rebolini, G. Bruno, F. Bocchi, M. Marelli, A. Orini: The new Turbigo–Rho 380 kV transmission line: an example of the use of underground XLPE cables in a meshed transmission grid, *Cigré 2006, Paper B1-301*
21. G. Bruno, E. Di Bartolomeo, M. Marani, M. Rebolini, R. Rendina, S. Tricoli: The 380 kV Turbigo (Rho)-Ospiate overhead-cable mixed line, *L'Energia Elettrica*, Vol. 84, No. 1, 2007, pp. 27–36 (in Italian)
22. Law 22 February 2001, n. 36: Framework Law on the protection against exposure to electric, magnetic, and electromagnetic fields, published on Official Gazette of the Italian Republic n. 55, March 7, 2001 (in Italian)
23. Decree of the President of the Council of Ministers, DPCM 8 July 2003: Establishment of exposure limits, attention values, and quality goal sto protect the population against power frequency (50 Hz) electric and magnetic fields generated by power lines, published on the Official Gazette of the Italian Republic n. 200, August 29, 2003 (in Italian)
24. CEI 11-60 (Italian Standard): Carrying capacity at thermic limit of overhead lines exceeding 100 kV, 2002

Chapter 2
The Positive Sequence Model of Symmetrical Lines

2.1 Introduction

In the power frequency steady state regimes of a power line, the positive sequence model is often assumed: this model is almost always acceptable for studies of balanced regimes of overhead lines (even if they are asymmetrical and with earth wires) whereas, in the case of single-core, unarmoured UGCs, it can be accepted only if cross-bonding with phase transpositions is performed in order to both zero the induced current circulation in the sheaths and symmetrize the system.

On the contrary for the Gas Insulated transmission Line the solid-bonding connection is required: the lowest value of the enclosure resistance allows the induced current in each enclosure to be almost equal in magnitude and opposite in angle to the corresponding phase current, so that a very good approximation consists in considering each phase as decoupled from the others (not only capacitively but also inductively), and hence in modelling the line at balanced regimes by means of the positive-sequence circuit.

After these remarks, it appears licit to turn to the classical treatment of the AC uniformly distributed parameter transmission lines.

However, it is worth remembering that also in the aforementioned cases, the determination of equivalent parameters to be introduced in the positive sequence model hold a given degree of uncertainty: the multiconductor analysis together with experimental results can be helpful to deepen, with more accuracy, the knowledge of the phenomena.

2.2 The Transmission Matrix of a Uniform Line

Let us consider a line at power frequency steady state regime as an uniform symmetrical transmission line: Figure 2.1 shows the single-phase model with distributed parameters at positive sequence, where each infinitesimal element $\mathrm{d}x$ is qualified by the four parameters (also called primary constants) r, ℓ, g, and c. Their values can

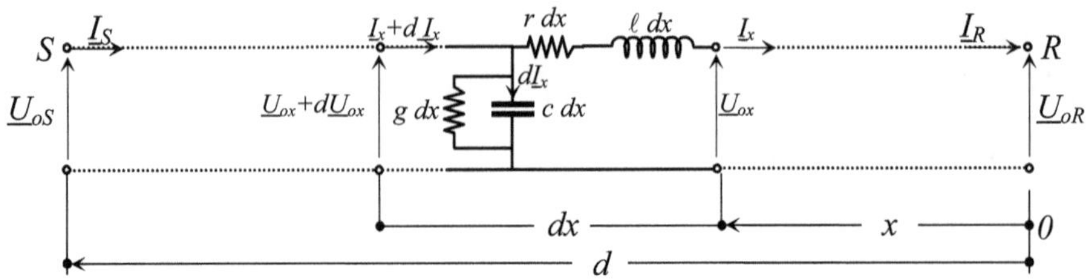

Figure 2.1 Single-phase model at positive sequence of a line

be computed after Sections 2.3.1–2.3.4 in case of cross-bonded cables with phase transpositions and after Sections 2.4.1–2.4.4 in case of solid-bonded GILs. For OHL the formulae are well-known.

The voltage and current references at the sending- and receiving-ends (both of the entire line and of the infinitesimal element) are also defined in Figure 2.1.

In the steady state regime each element of infinitesimal length dx has, at ω angular frequency, a longitudinal impedance and a shunt admittance given by:

$$\begin{aligned} \underline{z}\mathrm{d}x &= (r + \mathrm{j}\omega\ell)\mathrm{d}x \; , \\ \underline{y}\mathrm{d}x &= (g + \mathrm{j}\omega c)\mathrm{d}x \; ; \end{aligned} \tag{2.1}$$

where $\underline{z}$ is the per unit length (or kilometric) longitudinal impedance expressed in Ω/km and $\underline{y}$ the per unit length (or kilometric) shunt admittance in S/km.

The voltage increase $\mathrm{d}\underline{U}_{0x}$ between the two ports of the infinitesimal element dx (located at distance x [km] from the receiving-end R) is the voltage drop on the longitudinal impedance $\underline{z}\mathrm{d}x$ so that

$$\mathrm{d}\underline{U}_{0x} = \underline{I}_x \underline{z}\mathrm{d}x \; . \tag{2.2}$$

The current variation $\mathrm{d}\underline{I}_x$ between the two ports of the infinitesimal element is equal to the current derived from the infinitesimal shunt admittance

$$\mathrm{d}\underline{I}_x = [\underline{U}_{0x} + \mathrm{d}\underline{U}_{0x}] \cdot \underline{y} \cdot \mathrm{d}x \; . \tag{2.3}$$

By considering that in (2.3) the term of superior order $\mathrm{d}\underline{U}_{0x} \cdot \underline{y}\mathrm{d}x$ can be neglected, (2.2) and (2.3) can be rewritten in the following form:

$$\frac{\mathrm{d}\underline{U}_{0x}}{\mathrm{d}x} = \underline{z}\,\underline{I}_x \; , \tag{2.4}$$

$$\frac{\mathrm{d}\underline{I}_x}{\mathrm{d}x} = \underline{y}\underline{U}_{0x} \; . \tag{2.5}$$

It is worth remembering that the steady state phasors $\underline{U}_{0x}$ and $\underline{I}_x$, rotating with constant angular frequency ω determine in the time domain (with their projections on a reference axis depending upon their phase displacement) the sinusoidal instantaneous values. Consequently it remains only to solve the system of differential equations composed of (2.4) and (2.5) as if the current and voltage phasors were complex functions of the sole spatial variable x.

By $\mathcal{L}$-transforming ($\mathcal{L}$ = *Laplace*) the complex functions $\underline{U}_{0x} = \underline{U}(x)$ and $\underline{I}_x = \underline{I}(x)$ of the x real variable and remembering the well-known transform of a function's derivative i.e.

$$\mathcal{L}\left\{\frac{\mathrm{d}f(x)}{\mathrm{d}x}\right\} = s\mathcal{L}\{f(x)\} - f(0)$$

the following system arises

$$\begin{cases} s\mathcal{L}\{\underline{U}_{0x}\} - \underline{U}_{0R} = \underline{z} \cdot \mathcal{L}\{\underline{I}_x\}\,, \\ s\mathcal{L}\{\underline{I}_x\} - \underline{I}_R = \underline{y} \cdot \mathcal{L}\{\underline{U}_{0x}\}\,, \end{cases}$$

where (with the abscissa origin at receiving end) the obvious relations $\underline{U}_{0(x=0)} = \underline{U}_{0R}$ and $\underline{I}_{(x=0)} = \underline{I}_R$ have been applied.

Rearranging and reordering, it has

$$\begin{cases} s\mathcal{L}\{\underline{U}_{0x}\} - \underline{z}\mathcal{L}\{\underline{I}_x\} = \underline{U}_{0R}\,, \\ -\underline{y}\mathcal{L}\{\underline{U}_{0x}\} + s\mathcal{L}\{\underline{I}_x\} = \underline{I}_R\,, \end{cases}$$

from which it is immediate to obtain (by means of Cramer's rule)

$$\mathcal{L}\{\underline{U}_{0x}\} = \frac{s\underline{U}_{0R} + \underline{z}\,\underline{I}_R}{s^2 - \underline{z} \cdot \underline{y}}\,, \tag{2.6}$$

$$\mathcal{L}\{\underline{I}_x\} = \frac{s\underline{I}_R + \underline{y}\underline{U}_{0R}}{s^2 - \underline{z} \cdot \underline{y}}. \tag{2.7}$$

If the new quantities $\underline{k}$ and $\underline{Z}_c$ are defined

$$\underline{k} = \sqrt{\underline{z} \cdot \underline{y}}\,, \quad \underline{Z}_c = \sqrt{\frac{\underline{z}}{\underline{y}}}\,,$$

so that:

$$\underline{z} = \underline{k}\,\underline{Z}_c\,, \quad \underline{y} = \frac{\underline{k}}{\underline{Z}_c}\,,$$

it is possible to express (2.6) and (2.7) in their final form:

$$\mathcal{L}\{\underline{U}_{0x}\} = \frac{s\underline{U}_{0R} + \underline{k}\,\underline{Z}_c\underline{I}_R}{s^2 - \underline{k}^2}\,, \tag{2.8}$$

$$\mathcal{L}\{\underline{I}_x\} = \frac{s\underline{I}_R + \frac{\underline{k}}{\underline{Z}_c}\underline{U}_{0R}}{s^2 - \underline{k}^2}. \tag{2.9}$$

By implementing the inverse Laplace transform of them, the current and voltage phasors as a function of x are expressed by

$$\begin{cases} \underline{U}_{0x} = \underline{A}_x \cdot \underline{U}_{0R} + \underline{B}_x \cdot \underline{I}_R\,, & (2.10) \\ \underline{I}_x = \underline{C}_x \cdot \underline{U}_{0R} + \underline{D}_x \cdot \underline{I}_R\,, & (2.11) \end{cases}$$

where

$$\begin{aligned} \underline{A}_x &= \cosh(\underline{k}x)\ ; & \underline{B}_x &= \underline{Z}_c \cdot \sinh(\underline{k}x)\ ; \\ \underline{C}_x &= [\sinh(\underline{k}x)]/\underline{Z}_c\ ; & \underline{D}_x &= \cosh(\underline{k}x)\ , \end{aligned} \tag{2.12}$$

x	[km]	distance to the receiving end of the line;
$\underline{k} = \sqrt{\underline{z} \cdot \underline{y}}$	$[\mathrm{km}^{-1}]$	propagation factor;
$\underline{Z}_c = \sqrt{\frac{\underline{z}}{\underline{y}}}$	[Ω]	characteristic or natural impedance.

If the hyperbolic functions are expressed as exponential ones (by using the Euler–Lambert relations) it is possible to obtain the transmission formulae (solution of the telegraphist's equations) which were first written in 1876 by O. Heaviside [1] and then used, rearranged and divulged by Arnold [2], Rössler [3], Steinmetz [4] et al.

If $x = d$, it yields

$$\underline{U}_{0(x=d)} = \underline{U}_{0S} \quad \text{and} \quad \underline{I}_{x=d} = \underline{I}_S$$

so that

$$\underline{U}_{0S} = \underline{A} \cdot \underline{U}_{0R} + \underline{B} \cdot \underline{I}_R\ , \tag{2.13}$$

$$\underline{I}_S = \underline{C} \cdot \underline{U}_{0R} + \underline{D} \cdot \underline{I}_R\ , \tag{2.14}$$

where

$$\begin{aligned} \underline{A} &= \cosh(\underline{k}d)\ ; & \underline{B} &= \underline{Z}_c \cdot \sinh(\underline{k}d)\ ; \\ \underline{C} &= [\sinh(\underline{k}d)]/\underline{Z}_c\ ; & \underline{D} &= \cosh(\underline{k}d)\ . \end{aligned} \tag{2.15}$$

The single-phase model of the line (of length d) can therefore be considered as a two-port network which is characterized by means of $\underline{A}$, $\underline{B}$, $\underline{C}$, $\underline{D}$ parameters, also called as transmission hybrid parameters since they have different units (and different circuital meanings) as highlighted in any power system analysis book [5, 6].

Since the line has uniformly distributed parameters, the two-port network results symmetric so that:

$$\underline{A} = \underline{D}\ .$$

Equations 2.13 and 2.14 can be rearranged in the concise elegant matrix form, where $\underline{\boldsymbol{M}}$ is often named transmission matrix,

$$\underbrace{\begin{bmatrix} \underline{U}_{0S} \\ \underline{I}_S \end{bmatrix}}_{\underline{\boldsymbol{w}}_S} = \underbrace{\begin{bmatrix} \underline{A} & \underline{B} \\ \underline{C} & \underline{D} \end{bmatrix}}_{\underline{\boldsymbol{M}}} \cdot \underbrace{\begin{bmatrix} \underline{U}_{0R} \\ \underline{I}_R \end{bmatrix}}_{\underline{\boldsymbol{w}}_R} \tag{2.16}$$

which allows obtaining the hybrid vector $\underline{\boldsymbol{w}}_S$ at sending-end once fixed the hybrid vector $\underline{\boldsymbol{w}}_R$ at receiving-end.

The relations (2.15) allow verifying that

$$\det \underline{M} = \underline{A}\,\underline{D} - \underline{B}\,\underline{C} = [\cosh(kd)]^2 - [\sinh(kd)]^2 = 1 \,, \tag{2.17}$$

so confirming that the two-port network constituted by the line obviously satisfies the reciprocity principle.

2.3 Computation of Single-Core Cable Kilometric Parameters

As already stated and remarked in Sections 2.1 and 2.2, the computation of AC primary parameters is referred to a cross-bonded (with phase transpositions) single-core unarmoured cable.

2.3.1 *Computation of r (Cable)*

It is important to reaffirm that with the hypothesized perfect cross-bonding the induced sheath currents are zeroed and that, as is well-known, the alternating current Joule losses in the phase conductor are (due to the skin and proximity effects) higher than those in direct current.

This fact brings to consider for the phase conductor a kilometric resistance r_{ac} greater than that in direct current $r_{\text{dc}\vartheta^\circ}$ as shown in [7](IEC 60287):

$$r_{\text{ac}} = r_{\text{dc}\vartheta^\circ} \cdot (1 + y_{\text{s}} + y_{\text{p}}) \quad [\Omega/\text{km}] \,, \tag{2.18}$$

where:

y_{s} ⇒ skin effect factor;

y_{p} ⇒ proximity effect factor;

$r_{\text{dc}\vartheta^\circ}$ ⇒ DC resistance computed at the maximum operating conductor temperature ϑ°

r_{ac} ⇒ kilometric cable resistance at maximum operating temperature ϑ°.

Table 2.1 Permissible conductor temperatures for HV and EHV cables [8]

Cable insulating material	Maximum operating temperature [°C]	Short circuit temperature [°C]
Impregnated paper	85–90	160–180
Low-density polyethylene – LDPE	70	150
High-density polyethylene – HDPE	80	180
Cross-linked polyethylene – XLPE	90	250
Ethylene Propylen Rubber – EPR	90	250

Table 2.2 Electrical resistivity at 20 °C and temperature coefficients of materials used in cable construction [7]

Material used in	Resistivity ρ_{20° Ω m at 20 °C	Temperature coefficient α
Conductors		
Copper	1.7241×10^{-8}	3.93×10^{-3}
Aluminium	2.8264×10^{-8}	4.03×10^{-3}
Sheaths and armours		
Lead or lead alloy	21.4×10^{-8}	4.0×10^{-3}
Steel	13.8×10^{-8}	4.5×10^{-3}
Bronze	3.5×10^{-8}	3.0×10^{-3}
Stainless steel	70×10^{-8}	Negligible
Aluminium	2.84×10^{-8}	4.03×10^{-3}

Usually it is conservative to assume that ϑ° is the maximum operating conductor temperature [8] compatible with the insulating materials (see Table 2.1). It should be noted that XLPE and EPR are the most interesting dielectrics.

For the computation of $r_{\text{dc}\vartheta^\circ}$ the following well-known formula can be used starting from the kilometric resistance r_{20° at 20 °C

$$r_{\text{dc}\vartheta^\circ} = r_{20^\circ} \cdot [1 + \alpha \cdot (\vartheta^\circ - 20^\circ)] \quad [\Omega/\text{km}] \,, \tag{2.19}$$

where:

$\alpha \Rightarrow$ temperature coefficient 1/°C;

$\vartheta^\circ \Rightarrow$ maximum operating conductor temperature.

With regard to the resistivities and temperature coefficients of the metals used in cable construction, Table 2.2 sums up the values [7].

The computation of y_s and y_p must be performed after the detailed formulae of IEC 60287 [7].

With regard to the large cross section conductors, it is worth remembering a segmental conductor design (Milliken H. Electrical Cable. Patent Publication 1933, US 1904162), which is commonly referred to as a Milliken conductor: it has insulated segments consisted of layers of wires (more or less insulated each other) stranded around the segment "centre" aiming at reducing the magnitude of skin and proximity effects [9, 10].

Nowadays, cross sections up to 2500 mm^2 are customarily used and this topic continues to be reviewed [11] and developed so that any formula must be used with due caution (possibly supported by experimental validations of manufacturer).

2.3.2 Computation of ℓ (Cable)

In the hypothesis that the current distribution is uniform in the whole cross-section of the phase conductor and that the sheath currents are zeroed by the cross-bonding,

Figure 2.2 Single-core cable laying arrangements

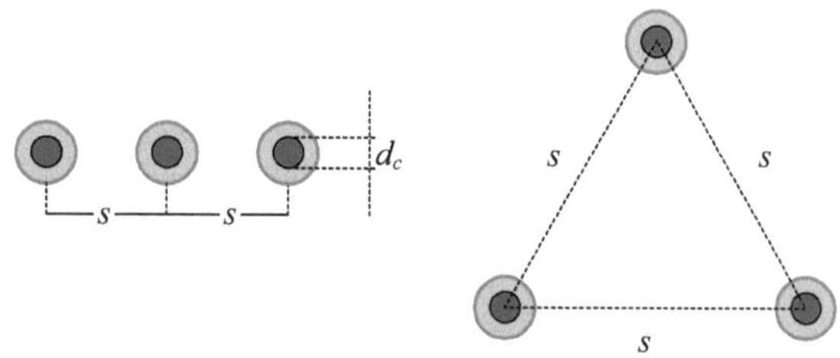

flat arrangement

Open trefoil arrangement with axial spacing s

the kilometric inductance (to be given to each phase) can be computed by means of

$$\ell = 0.2 \ln\left(\frac{K \cdot s}{GMR}\right) \quad [\text{mH/km}]\,, \tag{2.20}$$

where:

$s \Rightarrow$ axial spacing (mm); (see Figure 2.2);
$K \Rightarrow$ 1 for trefoil arrangement;
$K \Rightarrow$ 1.26 for flat laying (this value can be easily confirmed by means of the Geometrical Mean Distance among conductor centres $GMD = \sqrt[3]{s_{12} \cdot s_{23} \cdot s_{13}} = \sqrt[3]{s \cdot s \cdot 2s} = 1.26 \cdot s$ when the cross-bonding is performed with phase transpositions);
$GMR \Rightarrow$ geometrical mean radius (mm) of the phase conductor.

Usually in EHV XLPE-cables the conductor is hollow Milliken type; for *GMR*, see [12], Vol. II, page 155.

It is worth noting that the axial spacing can be reduced to a minimum (i.e. touching trefoil) but with a decreased ampacity.

2.3.3 Computation of c (Cable)

The values of the single-core cable kilometric capacitance c (between the phase conductor and the metallic sheath) can be computed by considering the cylindrical symmetry, so that (with $\varepsilon_0 = 8.8542 \cdot 10^{-12}\text{F/m}$)

$$(c) = \frac{2\pi\varepsilon_0\varepsilon_r}{\ln\left(\frac{d_0}{d_1}\right)} \quad [\text{F/m}] \Rightarrow c = \frac{\varepsilon_r}{18 \ln\left(\frac{d_0}{d_1}\right)} \quad [\mu\text{F/km}]\,, \tag{2.21}$$

where:

ε_r ⇒ relative dielectric constant of the insulating material;
d_0 ⇒ outer diameter of insulating medium excluding semi conductive screen or layer;
d_1 ⇒ conductor diameter including semi conductive screen or layer, if any.

2.3.4 Computation of g (Cable)

The shunt conductance g (S/km) is the parameter taking into account the active power "dielectric" losses $g \cdot U_0^2$ (per kilometre) given for each phase by the electric field in the insulating material [10, 13].

It is worth highlighting that the circuital element given to it (conductive shunting of capacity as in Figure 2.1) allows evaluating correctly the dielectric losses, but it not constitutes a realistic modelling in order to qualify the complex physical phenomena which are the fundamental cause and are referable to a kind of hysteresis in the dielectric polarization given by the alternating electric field at angular frequency ω: only in a very small part the losses are referable to a proper Joule effect due to imperfect insulation, also present in DC.

The shunt conductance can be computed with

$$g = b \cdot \tan\delta \ , \tag{2.22}$$

where

$\tan\delta$ ⇒ loss factor of insulating material (see Table 2.3),
$b = \omega c$ [S/km] ⇒ the per unit length capacitive susceptance of the cable,

so highlighting active power losses proportional to capacitive power absorbed along the cable and hence proportional to the frequency and to the square of the voltage.

The total (in the three phases) dielectric losses are given by

$$w_{\mathrm{dTOT}} = 3 \cdot g \cdot U_0^2 = g \cdot U_n^2 \quad [\mathrm{W/km}]$$

where U_n [V] is the rated phase-to-phase voltage (see Table 6.4).

The dielectric losses of HV and EHV cables ought to be always accounted for whilst they can be neglected for low and medium voltage since they are very small in comparison with the Joule losses in the longitudinal phase resistance.

In any case, the IEC standard 60287-1-1 states that the dielectric losses should be taken into account for values of the phase-to-ground voltage U_0 equal to or greater than those of the Table 2.4.

Table 2.3 Values of relative permittivity ε_r and loss factor *tan* δ at 50–60 Hz (see Section 2.3.4), for different insulating materials after [7]

Type of cable	ε_r	*tan* δ
Cables insulated with impregnated paper		
Solid-type, fully impregnated, preimpregnated or mass-impregnated nondraining	4	0.01
Oil-filled, self-contained		
up to $U_0 = 36$ kV	3.6	0.0035
up to $U_0 = 87$ kV	3.6	0.0033
up to $U_0 = 160$ kV	3.5	0.0030
up to $U_0 = 220$ kV	3.5	0.0028
Oil-pressure, pipe-type	3.7	0.0045
Internal gas-pressure	3.4	0.0045
External gas-pressure	3.6	0.0040
Cables with other types of insulation		
Butyl rubber	4	0.050
EPR – up to and including 18/30 kV	3	0.020
EPR – above 18/30 kV	3	0.005
PVC	8	0.1
PE (HD and LD)	2.3	0.001
XLPE up to and including 18/30(36) kV – unfilled	2.5	0.004
XLPE above 18/30(36) kV – unfilled	2.5	0.001
XLPE above 18/30(36) kV – filled	3	0.005
Paper–polypropylene–paper (PPL) $\geq$ 63/110 kV	2.8	0.0014

Table 2.4 Values of phase-to-ground voltage above (or equal to) which the dielectric losses should be considered [7]

Type of cable	U_0 [kV]
Cables insulated with impregnated paper	
Solid-type	38
Oil-filled and gas-pressure	63.5
Cables with other types of insulation	
Butyl rubber	18
EPR	63.5
PVC	6
PE (HD and LD)	127
XLPE (unfilled)	127
XLPE (filled)	63.5

2.4 Computation of GIL Kilometric Parameters

It has been already premised in Section 2.1 that the single-phase model at the positive sequence of a GIL with solid-bonding connection represents a good approximation: if the enclosures of a GIL (which are always very low resistive) are solid-bonded, they allow the circulation of induced currents almost equal (in magnitude)

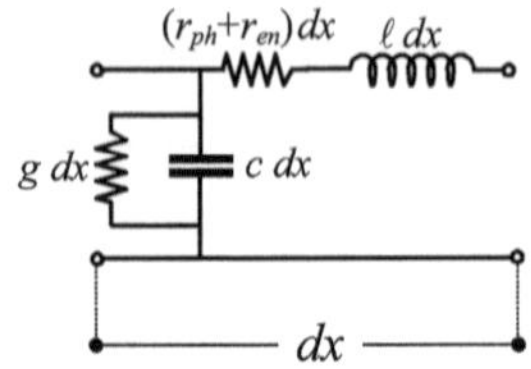

$r_{ph} \Rightarrow$ kilometric resistance of the phase conductor
$r_{en} \Rightarrow$ kilometric resistance of the enclosure
$\ell \Rightarrow$ kilometric inductance of the coaxial circuit
$c \Rightarrow$ kilometric capacitance of the coaxial circuit
$g \Rightarrow$ kilometric conductance of the coaxial circuit

Figure 2.3 Single-phase model of a GIL element dx at positive sequence

to those of phases and opposite in angle [14–17], so that they can be considered as return paths of each current in simple single phase circuits electromagnetically decoupled, having parameters computable on the basis of typical coaxial configurations (see Figure 2.3).

2.4.1 Computation of GIL Apparent Kilometric Resistance r

It is plain that, with the aforementioned hypothesis for the phase and enclosure current phasors, the total Joule losses must be computed by fixing (in the model of Figure 2.3) for the phase conductor an apparent kilometric resistance r equal to the sum of phase resistance r_{ph} and enclosure resistance r_{en}

$$r = r_{\text{ph}} + r_{\text{en}} \quad [\Omega/\text{km}] \,. \tag{2.23}$$

It is worth mentioning that the thicknesses of the pipes and the used spacing render negligible the skin and proximity effects at power frequency [16]: so that r_{ph} and r_{en} can be evaluated in DC. The value of phase resistance r_{ph} is given by:

$$r_{\text{ph}} = \frac{\rho_{\text{ph}}}{S_{\text{ph}}} \quad [\Omega/\text{km}] \,, \tag{2.24}$$

where ρ_{ph} is expressed in [$\Omega\,\text{mm}^2/\text{km}$] and the cross-sectional area is

$$S_{\text{ph}} = \pi \cdot \left(R_2^2 - R_1^2\right) \quad [\text{mm}^2] \tag{2.25}$$

with R_2 and R_1 the outer and inner radii of phase conductor expressed in millimetres (see Figure 2.4).

Data concerning the involved materials and the resistivities can be obtained by Tables 3.4 and 3.5 or by more precise "manufacturer's" information.

Analogously for the computation of r_{en}:

$$r_{\text{en}} = \frac{\rho_{\text{en}}}{S_{\text{en}}} \quad [\Omega/\text{km}] \,, \tag{2.26}$$

$$S_{\text{en}} = \pi \cdot \left(R_4^2 - R_3^2\right) \quad [\text{mm}^2] \,, \tag{2.27}$$

where R_4 and R_3 are the outer and inner radii of the enclosure expressed in mm.

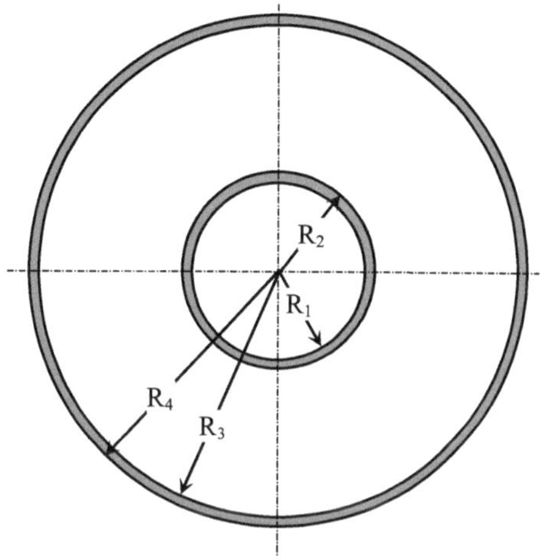

Figure 2.4 One phase and corresponding enclosure of a GIL

The resistivities ρ_{ph} and ρ_{en} of material ($\Omega\,\text{mm}^2/\text{km}$) must be evaluated at the operating temperature: the maximum allowable ones are stated in IEC 61640 [18].

2.4.2 Computation of GIL Kilometric Inductance ℓ

By considering the typical coaxial structure of a GIL (see sketch in Figure 2.4) and by adopting the above-mentioned solid-bonding technique, if it is assumed that opposite currents with magnitude I flow in the phase conductor and in the enclosure with uniform distribution (no skin and proximity effects), it is possible to compute the kilometric inductance as the sum of three terms $\ell = \ell_a + \ell_b + \ell_c$ (the first represents the phase internal inductance, the second the inductance due to the magnetic field between phase and enclosure, the third the enclosure internal inductance); it is also simple to ascertain that the value of ℓ_b is prevalent among the others and that, for the assumed hypothesis, the magnetic field external to the enclosure is zeroed.

Since (see Figure 2.4) R_1, R_2 are the phase inner and outer radii respectively and R_3, R_4 the enclosure inner and outer radii respectively, it is easy, by integrating the magnetic field between R_2 and R_3, to express the inductance l_b through

$$\ell_b = \frac{\mu_0}{2\pi} \ln \frac{R_3}{R_2} \quad [\text{H/m}]\,, \tag{2.28}$$

where:

$\mu_0 \quad \Rightarrow \quad$ magnetic permeability of the free space $= 4\pi \times 10^{-7}$ H/m.

In the usual installations, it is $\ln\left(\frac{R_3}{R_2}\right) \cong 1$ in order to obtain, as is well-known, an optimal configuration of the electric field.

In order to obtain the expression of ℓ_a it is convenient to premise that (in accordance with electromagnetic theory) the following relation can be written

$$en_a = \int \left[\left(\mu_0 H^2\right)/2\right] \mathrm{d}v = \ell_a I^2/2\,; \tag{2.29}$$

it correlates the magnetic energy en_a, expressed by the integral computed in the whole volume of the phase conductor (of 1 m length), with the corresponding quantity $\ell_a I^2/2$.

Once computed en_a, it yields immediately

$$\ell_a = 2en_a/I^2\,. \tag{2.30}$$

It is sufficient to observe that the magnitude of H_x in the integration interval R_1–R_2 can be written (since the circular symmetry and the hypothesis of uniform current density σ_{ph} in the phase) by means of

$$H_x = \frac{\sigma_{ph}\cdot\pi\cdot\left(x^2-R_1^2\right)}{2\pi x}\,;\quad \sigma_{ph} = \frac{I}{\pi\left(R_2^2-R_1^2\right)}\quad (R_1 \le x \le R_2)\,.$$

Analogously, the computation of ℓ_c can be performed by writing for the magnitude of H_x in the integration interval R_3–R_4 the two relations (note that the current $-I$ flows in the enclosure)

$$H_x = \frac{I}{2\pi x} + \frac{\sigma_{en}\cdot\pi\cdot\left(x^2-R_3^2\right)}{2\pi x}\,;\quad \sigma_{en} = -\frac{I}{\pi\left(R_4^2-R_3^2\right)}\quad (R_3 \le x \le R_4)\,.$$

By developing the above-mentioned procedure, the following formulae (see also [19, 20]) can be achieved

$$\ell_a = \frac{\mu_0}{2\cdot\pi}\cdot\left[\frac{R_1^4}{\left(R_2^2-R_1^2\right)^2}\ln\left(\frac{R_2}{R_1}\right) + \frac{\left(R_2^2-3R_1^2\right)}{4\left(R_2^2-R_1^2\right)}\right]\quad (\mathrm{H/m})\,, \tag{2.31}$$

$$\ell_c = \frac{\mu_0}{2\cdot\pi}\cdot\left[\frac{R_4^4}{\left(R_4^2-R_3^2\right)^2}\ln\left(\frac{R_4}{R_3}\right) + \frac{\left(R_3^2-3R_4^2\right)}{4\left(R_4^2-R_3^2\right)}\right]\quad (\mathrm{H/m})\,. \tag{2.32}$$

In order to have some orders of magnitude, it seems useful this example. Once the following values are assumed:

$$R_1 = 80\,\mathrm{mm}\,,\quad R_2 = 90\,\mathrm{mm}\,,\quad R_3 = 250\,\mathrm{mm}\,,\quad R_4 = 260\,\mathrm{mm}\,,$$

by means of (2.31), (2.28) and (2.32) and by converting from H/m to mH/km the following values yield:

$$\ell_a = 0.0074\,,\quad \ell_b = 0.20433\,,\quad \ell_c = 0.0026\quad \mathrm{mH/km}\,,$$
$$\ell = \ell_a + \ell_b + \ell_c = 0.2144\,\mathrm{mH/km}\,,$$

where it appears that ℓ_b is the most important one and, as first approach, it always gives sufficient accuracy.

2.4.3 The Computation of GIL Kilometric Capacitance c

The value of capacitance can be easily computed considering the cylindrical symmetry and that the permittivity of insulating gas mixture is $\varepsilon_0 = 8.8542\times 10^{-12}\,\mathrm{F/m}$ (as for the free space):

$$(c) = \frac{2\pi\varepsilon_0}{\ln\left(\frac{R_3}{R_2}\right)} \quad [\mathrm{F/m}] \quad \Rightarrow \quad c = \frac{1}{18\cdot\ln\left(\frac{R_3}{R_2}\right)} \quad [\mu\mathrm{F/km}]\,. \tag{2.33}$$

2.4.4 Computation of GIL Kilometric Shunt Conductance g

It is worth noting that the dielectric used in the GIL (pure pressurized SF_6 or a mixture of it with N_2, at about 7 bar for $U = 380\,\mathrm{kV}$) has always a very low loss factor and gives losses that result absolutely negligible also if compared with the small Joule longitudinal losses due to the low apparent resistance r expressed by (2.23).

Also in precise economic evaluations of losses, it is always assumed $g = 0$.

2.5 Some Other Matrix Relations Deriving from the Fundamental One

In the transmission equations (2.13) and (2.14), there are four variables: the voltage and current phasors at sending and receiving ends.

When two of the four variables are fixed, the remaining ones are univocally determined.

Consequently, in this paragraph, different ways of using the transmission equations are highlighted in order to both give the reader a test-bed for getting some practice and have suitable formulae for Chapters 3 and 4.

As a first example (2.34), as much typical as (2.16), can be derived by rearranging (2.13) and (2.14): it expresses the vector $\underline{w}_R$ (hybrid), composed of the voltage and current at receiving-end, as a function of the vector $\underline{w}_S$ at sending-end:

$$\underbrace{\begin{bmatrix} \underline{U}_{0R} \\ \underline{I}_R \end{bmatrix}}_{\underline{w}_R} = \underbrace{\begin{bmatrix} \underline{D} & -\underline{B} \\ -\underline{C} & \underline{A} \end{bmatrix}}_{\underline{M}_a} \cdot \underbrace{\begin{bmatrix} \underline{U}_{0S} \\ \underline{I}_S \end{bmatrix}}_{\underline{w}_S}; \quad \underline{M}_a = \underline{M}^{-1}\,. \tag{2.34}$$

It is easy to demonstrate that $\underline{M}_a$ is the inverse of $\underline{M}$ (or that the product of $\underline{M}$ and $\underline{M}_a$ is the identity matrix), by remembering that $\underline{A}\,\underline{D} - \underline{B}\,\underline{C} = 1$ in virtue of the reciprocity principle (see (2.17)).

As a second example, if the sending-end voltage phasor and the receiving-end current phasor are fixed, it is easy to obtain the matrix relation (with $\underline{D} \equiv \underline{A}$)

$$\begin{bmatrix} \underline{U}_{0R} \\ \underline{I}_S \end{bmatrix} = \underbrace{\begin{bmatrix} 1/\underline{A} & -\underline{B}/\underline{A} \\ \underline{C}/\underline{A} & 1/\underline{A} \end{bmatrix}}_{\underline{M}_b} \cdot \begin{bmatrix} \underline{U}_{0S} \\ \underline{I}_R \end{bmatrix} \tag{2.35}$$

which allows determining the other two phasors $\underline{U}_{0R}$ and $\underline{I}_S$.

As it will be used in Chapter 3, in order to know the variables along the transmission line, it is rather immediate to achieve from (2.34):

$$\underbrace{\begin{bmatrix} \underline{U}_{0x} \\ \underline{I}_x \end{bmatrix}}_{\underline{\boldsymbol{w}}_x} = \underbrace{\begin{bmatrix} \underline{D}_x & -\underline{B}_x \\ -\underline{C}_x & \underline{A}_x \end{bmatrix}}_{\underline{\boldsymbol{M}}_x} \cdot \underbrace{\begin{bmatrix} \underline{U}_{0S} \\ \underline{I}_S \end{bmatrix}}_{\underline{\boldsymbol{w}}_S} \tag{2.36}$$

which allows calculating the vector $\underline{\boldsymbol{w}}_x$ along the cable (at a distance x from S) having $\underline{A}_x \dots \underline{D}^x$ obvious meaning (once known the vector $\underline{\boldsymbol{w}}_S$).

It will be often sufficient to compute only the values of voltage $\underline{U}_{0T}$ and current $\underline{I}_T$ at line midpoint T with (2.37)

$$\underbrace{\begin{bmatrix} \underline{U}_{0T} \\ \underline{I}_T \end{bmatrix}}_{\underline{\boldsymbol{w}}_T} = \underbrace{\begin{bmatrix} \underline{D}_T & -\underline{B}_T \\ -\underline{C}_T & \underline{A}_T \end{bmatrix}}_{\underline{\boldsymbol{M}}_T} \cdot \underbrace{\begin{bmatrix} \underline{U}_{0S} \\ \underline{I}_S \end{bmatrix}}_{\underline{\boldsymbol{w}}_S} \tag{2.37}$$

by giving x the distance of T from S.

In Chapter 4 other relations, besides those of this chapter, will be developed bringing again reasons to appreciate the matrix algorithms [21, 22].

Furthermore, it is worth remembering the matrix relation:

$$\begin{bmatrix} \underline{U}_{0S} \\ \underline{U}_{0R} \end{bmatrix} = \underbrace{\begin{bmatrix} \underline{A}/\underline{C} & -1/\underline{C} \\ 1/\underline{C} & -\underline{D}/\underline{C} \end{bmatrix}}_{\underline{\boldsymbol{Z}}} \cdot \begin{bmatrix} \underline{I}_S \\ \underline{I}_R \end{bmatrix} \tag{2.38}$$

which involves an impedance matrix $\underline{\boldsymbol{Z}}$ (where for the current $\underline{I}_R$ at port R as well the same phasor references of Figure 2.1 are held).

In order to deduce the matrix relation in (2.38), (2.16) can be made explicit so that, by setting $\underline{I}_R = 0$:

$$\frac{\underline{U}_S}{\underline{I}_S} = \frac{\underline{A}}{\underline{C}} = \underline{Z}_{SS} \quad \text{and} \quad \frac{\underline{U}_R}{\underline{I}_S} = \frac{1}{\underline{C}} = \underline{Z}_{RS} \,.$$

By making (2.34) explicit, with $\underline{I}_S = 0$, it easily yields:

$$\frac{\underline{U}_R}{\underline{I}_R} = -\frac{\underline{D}}{\underline{C}} = \underline{Z}_{RR} \quad \text{and} \quad \frac{\underline{U}_S}{\underline{I}_R} = -\frac{1}{\underline{C}} = \underline{Z}_{SR} \,.$$

2.6 Cascade Connections of Two Port Networks (TPN)

Let us suppose to have the cascade connections of Figures 2.5 and 2.6.

As is well-known, the resulting transmission matrix is given by the product of the transmission matrices.

Figure 2.5 Cascade of 2 TPN

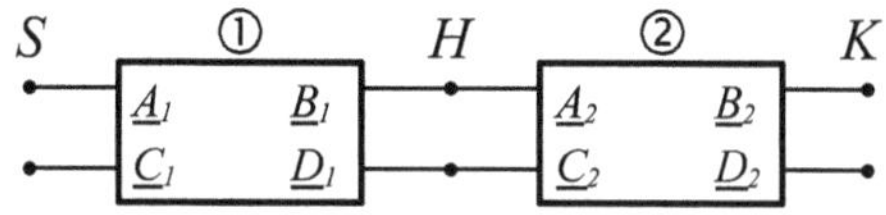

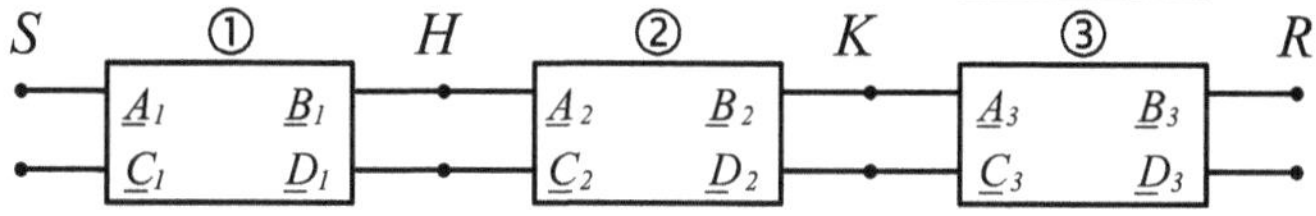

Figure 2.6 Cascade of 3 TPN

For each section ①, ② and ③ the corresponding transmission matrices can be calculated as already thoroughly explained in the previous sections:

$$\underbrace{\begin{array}{|c|c|}\hline \underline{A}_1 & \underline{B}_1 \\ \hline \underline{C}_1 & \underline{D}_1 \\ \hline \end{array}}_{\underline{\boldsymbol{M}}_1} ; \quad \underbrace{\begin{array}{|c|c|}\hline \underline{A}_2 & \underline{B}_2 \\ \hline \underline{C}_2 & \underline{D}_2 \\ \hline \end{array}}_{\underline{\boldsymbol{M}}_2} ; \quad \underbrace{\begin{array}{|c|c|}\hline \underline{A}_3 & \underline{B}_3 \\ \hline \underline{C}_3 & \underline{D}_3 \\ \hline \end{array}}_{\underline{\boldsymbol{M}}_3} . \tag{2.39}$$

The equivalent cascade matrices are given (with evident symbols)

$$\underline{\boldsymbol{M}}_{12} = \underline{\boldsymbol{M}}_1 \underline{\boldsymbol{M}}_2 = \begin{array}{|c|c|}\hline \underline{A}_{12} & \underline{B}_{12} \\ \hline \underline{C}_{12} & \underline{D}_{12} \\ \hline \end{array} ; \quad \underline{\boldsymbol{M}}_{123} = \underline{\boldsymbol{M}}_1 \underline{\boldsymbol{M}}_2 \underline{\boldsymbol{M}}_3 = \begin{array}{|c|c|}\hline \underline{A}_{123} & \underline{B}_{123} \\ \hline \underline{C}_{123} & \underline{D}_{123} \\ \hline \end{array} . \tag{2.40}$$

For example, by considering the ports in Figure 2.6, the following expressions appear immediately understandable:

$$\begin{aligned} \underline{\boldsymbol{w}}_S &= \underline{\boldsymbol{M}}_1 \cdot \underline{\boldsymbol{w}}_H ; & \qquad \underline{\boldsymbol{w}}_R &= \underline{\boldsymbol{M}}_3^{-1} \cdot \underline{\boldsymbol{w}}_K ; \\ \underline{\boldsymbol{w}}_S &= \underline{\boldsymbol{M}}_{12} \cdot \underline{\boldsymbol{w}}_K ; & \qquad \underline{\boldsymbol{w}}_H &= \underline{\boldsymbol{M}}_{23} \cdot \underline{\boldsymbol{w}}_R ; \\ \underline{\boldsymbol{w}}_S &= \underline{\boldsymbol{M}}_{123} \cdot \underline{\boldsymbol{w}}_R ; & \qquad \underline{\boldsymbol{w}}_R &= \underline{\boldsymbol{M}}_{123}^{-1} \cdot \underline{\boldsymbol{w}}_S . \end{aligned}$$

It is worth noting that generally even if $\underline{D}_1 \equiv \underline{A}_1$, $\underline{D}_2 \equiv \underline{A}_2$, $\underline{D}_3 \equiv \underline{A}_3$ (see (2.15)), the same is not necessarily valid in a cascade matrix where for example it is (but in particular cases) $\underline{D}_{12} \neq \underline{A}_{12}$ and $\underline{D}_{123} \neq \underline{A}_{123}$.

In any case, as long as non-reciprocal two port circuits are introduced (e.g. a phase shifter two-port), relations such as $\underline{A}_{123} \cdot \underline{D}_{123} - \underline{B}_{123} \cdot \underline{C}_{123} = 1$ remain still valid.

2.7 Parallel Connection of Equal Two Port Circuits Thermally and Electrically Decoupled

In case of a transmission link of a double circuit of equal cables in electric parallel (a situation that can be useful to gain a great power capability and enhance the reliability at the same time), the equivalent matrix can be easily deduced on the

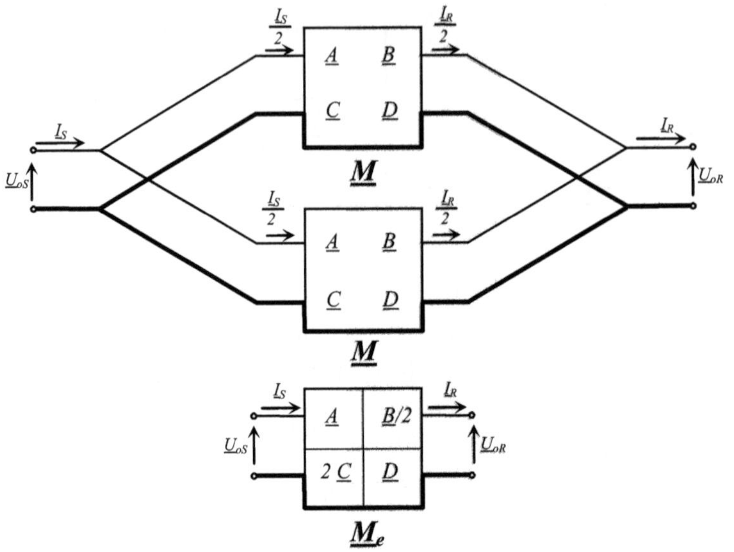

Figure 2.7 The parallel of two, perfectly decoupled, equal circuits

basis of the obvious equations in (2.41) valid for each of the two circuits (which are supposed to be thermally and electrically decoupled) with equal sharing of the currents shown in the scheme of Figure 2.7. Of course, it is true if the port voltages have the same reference (thick links in the sketch) as it is always valid in the positive and negative sequence models.

Therefore it is possible to determine the equations in (2.42) and hence the equivalent transmission matrix $\underline{\boldsymbol{M}}_e$ of the double circuit, once computed the single circuit matrix $\underline{\boldsymbol{M}}$,

$$\begin{Bmatrix} \underline{U}_{0S} = (\underline{A}) \cdot \underline{U}_{0R} + (\underline{B}) \cdot \frac{\underline{I}_R}{2} \, , \\ \frac{\underline{I}_S}{2} = (\underline{C}) \cdot \underline{U}_{0R} + (\underline{D}) \cdot \frac{\underline{I}_R}{2} \, , \end{Bmatrix} \tag{2.41}$$

$$\begin{Bmatrix} \underline{U}_{0S} = (\underline{A}) \cdot \underline{U}_{0R} + \left(\frac{\underline{B}}{2}\right) \cdot \underline{I}_R \, , \\ \underline{I}_S = (2\underline{C}) \cdot \underline{U}_{0R} + (\underline{D}) \cdot \underline{I}_R \, . \end{Bmatrix} \tag{2.42}$$

$$\underline{\boldsymbol{M}} = \begin{array}{|c|c|} \hline \underline{A} & \underline{B} \\ \hline \underline{C} & \underline{D} \\ \hline \end{array} \; ; \quad \underline{\boldsymbol{M}}_e = \begin{array}{|c|c|} \hline \underline{A} & \underline{B}/2 \\ \hline 2\underline{C} & \underline{D} \\ \hline \end{array} \, .$$

It seems interesting to note that the same result can be achieved (following a more intuitive procedure) by giving the double circuit the kilometric parameters $\underline{z}/2$ and $2\underline{y}$ (since $\underline{z}$ and $\underline{y}$ are those of the single circuit) so having the same propagation factor and the halving of the characteristic impedance.

2.8 The Shunt Reactive Compensation

It is known that the kilometric cable capacitive susceptance has often a so high value to determine (chiefly with great lengths, voltages and cross-sections) steady state and transient regimes which could become very severe or even unacceptable (as it will be shown in Chapters 3 and 4), unless a suitable shunt compensation is adopted.

Usually the cable link modelling must be completed with the insertion of the necessary single-phase shunt reactors at the two cable ends and in case (depending upon the link length) also in some other intermediate locations (lumped compensation).

By considering for example the circuital scheme of Figure 2.8, the steady state regime at angular frequency ω of a short link compensated only at the ends can be studied by means of the matrix cascade $\underline{\boldsymbol{M}}_{\xi} \cdot \underline{\boldsymbol{M}} \cdot \underline{\boldsymbol{M}}_{\xi}$, being that $\underline{\boldsymbol{M}}$ is the cable transmission matrix and $\underline{\boldsymbol{M}}_{\xi}$ is the transmission matrix which qualifies the behaviour of a shunt reactor with inductive susceptance equal to $\underline{Y}_{\xi}$ (as it will be shown in Section 2.8.2); it remains a useful basis also for the switching transients studies, since the subtransient regime at the same angular frequency ω (e.g. Sections 3.7 and 4.7) has a precise "indicial" role for the transient evaluations.

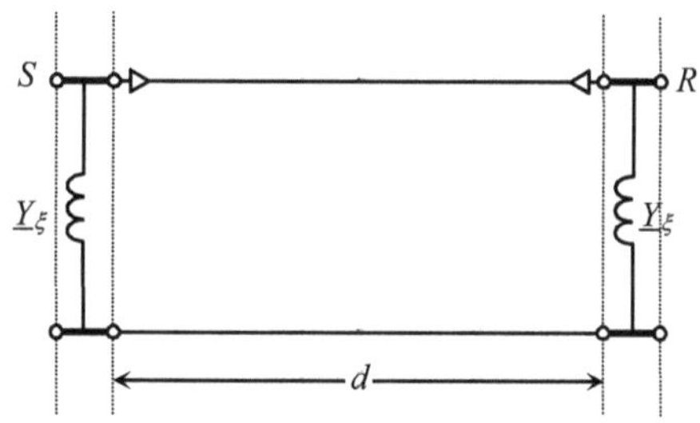

Figure 2.8 Lumped shunt compensation of short UGC (d < 20–25 km)

By considering that the lumped shunt reactors (located at short mutual distances < 20–25 km) can be numerous and can render the diagrams and the analytical relations heavy, it appears to be more efficient and meaningful, as a first approach, to imagine alternatively a uniformly distributed shunt compensation (see Section 2.8.1) in a way consistent with the uniformly distribution of the cable capacitance: it allows easily computing (and with sufficient accuracy) the necessary shunt compensation degree and having an equivalent transmission matrix which, even if ideal, is extremely useful.

2.8.1 The Uniformly Distributed Compensation

The compensation of the uniformly distributed cable capacitive susceptance would theoretically require the insertion of uniformly distributed inductive susceptance.

Since this is hard to imagine constructively, it is possible to think to insert, in each phase and at each kilometre, a reactor of inductive susceptance of finite value

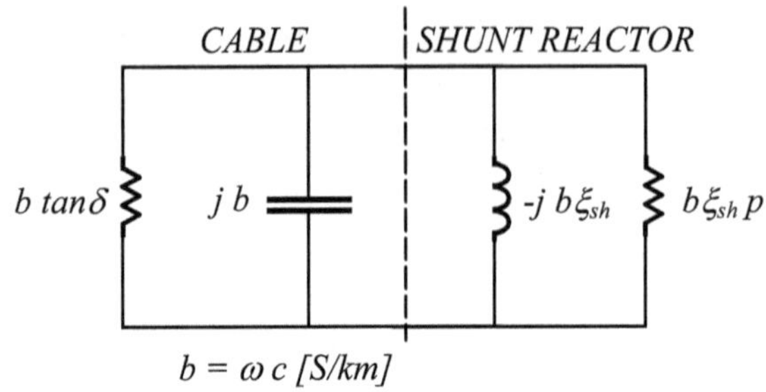

Figure 2.9 Kilometric shunt branch in the hypothesis of uniformly distributed shunt compensation

equal to $-\mathrm{j}\,\xi_{\mathrm{sh}}\omega c = -\mathrm{j}\,\xi_{\mathrm{sh}} b$ (see Figure 2.9), which compensates at each kilometre a fraction ξ_{sh} of the cable capacitive susceptance, bringing to compute the resulting kilometric susceptance $\underline{y}$ after (2.43)

$$\underline{y} = b \cdot \tan\delta + b \cdot \xi_{\mathrm{sh}} \cdot p + \mathrm{j}\, b \cdot (1 - \xi_{\mathrm{sh}})\;; \tag{2.43}$$

where the reactor "loss factor" p (which allows considering the reactor power losses, always very low) has been introduced.

In case of absence of compensation it has $\xi_{\mathrm{sh}} = 0$, and (2.43) returns the usual one:

$$\underline{y} = b \cdot (\tan\delta + \mathrm{j})\,. \tag{2.44}$$

Therefore, once fixed the shunt compensation degree ξ_{sh} and evaluated the suitable kilometric admittance after (2.43), it is possible to continue directly, by means of the already detailed relations, with the computation of all the new transmission parameters ($\underline{k}$, $\underline{Z}_{\mathrm{c}}$, and so on) up to the uniformly compensated link transmission matrix.

2.8.2 The Lumped Compensation

The transmission matrix obtained for a cable link uniformly compensated (as described in Section 2.8.1) allows developing (as it will be shown) quick procedures for a precise choice of the compensation degree ξ_{sh}: once fixed such value the link planning must be completed with an accurate configuration of the corresponding lumped compensation and with the check of it by means of a specific circuital analysis.

In the example of Figure 2.10, it is evident that the analysis regards the cascade of the three matrices $\underline{\boldsymbol{M}_{\xi}} \cdot \underline{\boldsymbol{M}} \cdot \underline{\boldsymbol{M}_{\xi}}$: it will be sufficient to form each matrix $\boldsymbol{M}_{\xi}$ (regarding a reactor of inductive admittance $\underline{Y_{\xi}}$) after (2.45):

$$\underline{\boldsymbol{M}_{\xi}} = \begin{array}{|c|c|}\hline 1 & 0 \\ \hline \underline{Y_{\xi}} & 1 \\ \hline \end{array} \tag{2.45}$$

which has an immediate confirmation in the circuit and in the obvious relations of Figure 2.11.

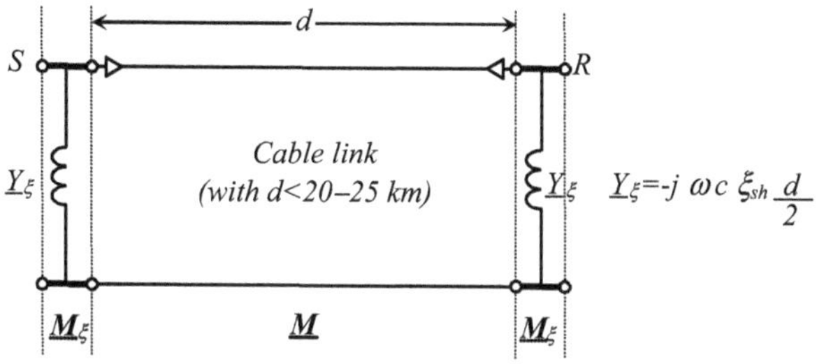

Figure 2.10 Cable link compensated at both ends

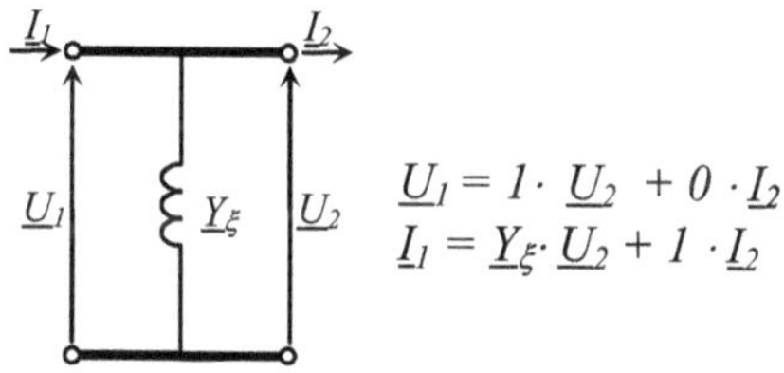

Figure 2.11 Shunt reactor circuit

If the cable to be compensated has a length greater than 20–25 km it is suitable to install shunt reactors also in n intermediate sections, respecting the limits shown in Figure 2.12.

The admittance value are subsequently modified in $\underline{Y}'_\xi = 2\underline{Y}_\xi = -\mathrm{j}\,\omega c \xi_{\mathrm{sh}} d/n$; it is almost trivial to observe that each cable section of length d/n must be modelled by its transmission matrix without compensation.

Once formed the cascade of the different two port circuits, the regime at angular frequency ω at each port can be analysed by making suitable use of the matricial algorithms already described in this chapter and of the procedures which will be exposed in the Chapters 3 and 4.

Hence it will be possible not only to ascertain the effectiveness of the hypothesis of distributed compensation, but also to evaluate in detail the effects, more or less positive, due to possible alternative schemes.

Usually the reactor percentage losses range between 0.1% and 0.3%.

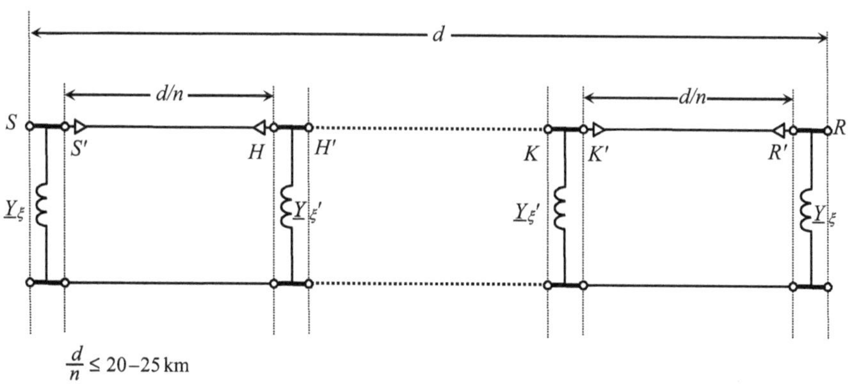

Figure 2.12 Lumped compensation of a cable link

References

1. O. Heaviside: Electrical papers, 2 Vol., 2nd Edition, Chelsea Publishing Company, New York, 1970
2. E. Arnold, A. Fraenckel, J.L. La Cour: Die Wechselstromtechnik, 5 Vol., Springer, Berlin, 1904–1914
3. G. Rössler: Die Fernleitung von Wechselströmen, Springer, Berlin, 1905
4. C.P. Steinmetz: Engineering Mathematics, McGraw-Hill, New York, 1910
5. J. Grainger, W. Stevenson: Power System Analysis, McGraw-Hill Science, 1994
6. C.A. Gross: Power system analysis, John Wiley & Sons, New York, 1986
7. IEC 60287: Electric cables – Calculation of the current rating (in 8 parts: 1.1, 1.2, 1.3, 2.1, 2.2, 3.1, 3.2, 3.3)
8. E. Peschke, E. von Olshausen: Cable systems for high and extra-high voltage, 1999, Publicis MDC Verlag, Germany
9. Cigré Technical Brochure No 272, Working Group B1.03: Large cross-sections and composite screens design, June 2005
10. G.J. Anders: Rating of electric power cables, IEEE PRESS, New York, 1997
11. D. Dubois, P. Mirebeau: The use of insulated wires Milliken conductors in high voltage power transmission underground AC lines, *Proc. of JICABLE 2007, Paper B8.1*, pp. 488–493
12. CCITT: Directives concerning the protection of telecommunication lines against harmful effects from electric power and electrified railway lines, Geneva, 1989
13. S.Y. King, N.A. Halfter: Underground power cables, Longman, London and New York, 1982
14. R. Benato, L. Fellin, A. Paolucci: Gas insulated lines: an alternative for bulk power transmission, *L'Energia Elettrica*, Vol. 77, No. 4/5, July–October 2000, pp. 60–63 (in Italian)
15. R. Benato, C. Di Mario, H. Koch: High capability applications of Long Gas Insulated Lines in Structures, *IEEE Trans. on Power Delivery*, Vol. 22, Issue 1, January 2007, pp. 619–626
16. R. Benato, F. Dughiero, M. Forzan, A. Paolucci: Proximity Effect and Magnetic Field Calculation in GIL and in Isolated Phase Bus Ducts, *IEEE Trans. on Magnetics*, Vol. 38, No. 2, March 2002, pp. 781–784
17. R. Benato, F. Dughiero: Solution of Coupled Electromagnetic and Thermal Problems in Gas Insulated Transmission Lines, *IEEE Trans. on Magnetics*, Vol. 39, No 3, May 2003, pp. 1741–1744
18. IEC report 61640: Rigid high-voltage, gas-insulated transmission lines for rated voltage of 72.5 kV and above, First edition, 1998
19. F.W. Grover: Inductance calculations; working formulas and tables, D. van Nostrand Co., New York, 1946
20. C. Snow: Formulas for computing capacitance and inductance, National Bureau of Standards Circular 544, Washington, 1954
21. R. Braae: Matrix algebra for Electrical Engineers. Sir Isaac Pitman & Sons LTD, London, 1963
22. G. Quilico: The matrices in electrotechnics, *L'Elettrotecnica*, Vol. XXIX in seven parts (25 September; 10 October; 25 October; 10 November; 25 November, 10 December, 25 December) 1942 (in Italian)

Chapter 3
Operating Capability of Long AC EHV Power Cables

3.1 Introduction

The operating constraints, regarding current and voltage regimes, involved in power transmission of long AC cable links are presented and here throughout developed. The possibility of transmitting high power rating is examined and the transmission length limits are evaluated. The procedure, which considers positive sequence models (as in Chapter 2), is essentially based on the classical transmission equations and their pertaining diagrams.

This chapter has been particularly devoted to XLPE cables and gas insulated lines, but the procedure is worth applying to any distributed-parameter transmission line included overhead lines. The capability charts are a useful tool to highlight the operating characteristics of a cable link and a guide to evaluate the results of different degrees of reactive compensation. The results can offer the transmission system operators very effective means in order to evaluate the existing and future underground cable links to achieve high performances and a long lifetime [1].

3.2 The Basic Constraints

It is well known that the lifetime of an AC cable with solid insulation strictly depends upon the story of the thermal (temperatures and their durations) and dielectric stresses suffered by the insulation. Therefore, a long service life can be reached with these constraints in any point along the cable:

- avoiding that the phase currents overcome the cable steady state ampacity I_a (even if, in many cases, the cyclic and emergency current regimes occur) meant as the continuous constant current just sufficient to produce asymptotically the maximum conductor temperature with acceptable dielectric degradation;
- avoiding that the phase-to-earth steady state voltages $|\underline{U}_0|$ overcome the highest RMS Voltage $U_m/\sqrt{3}$ (see the Insulation Co-ordination International Standard which also reports other limits for overvoltages of different duration) for cable both on-load and no-load conditions.

It is evident that foreseeing any conditions (steady state, transient and cyclic regimes) which the cable is subjected to during the operation of the network is rather difficult, if not after exhaustive power flow studies and transient simulations; notwithstanding, in order to frame in a preliminary way the possible performances, reasonable conservative conditions are generally:

- the power frequency currents at both ends (S for *sending-end* and R for *receiving-end*), during operation, must not exceed a current magnitude $I_c \leq I_a$;
- the phase-to-earth voltage level in one of the two ends must be fixed (in this book S will be always chosen).

Therefore, the following "basic constraints" are fixed:

$$|\underline{I}_R| \leq I_c \,, \tag{3.1}$$

$$|\underline{I}_S| \leq I_c \,, \tag{3.2}$$

$$|\underline{U}_{0S}| = U_{0c} \,. \tag{3.3}$$

In the following investigations, where $U_m = 420\,\text{kV}$, the level $U_{0c} = 230\,\text{kV} = 95\%\, U_m/\sqrt{3}$ has been always assumed.

A thorough analysis of the cable operating characteristics with the aforementioned constraints has been presented in [2], which constitutes (chiefly by means of a general abacus) one of the first elegant (dated 1986) approaches: it has been a strong suggestion for the authors.

The cable steady state ampacities in undergrounded installations are evaluated according to the thermal study affected by:

- Joule losses in the core conductor and the sheath (and armour if any);
- dielectric losses;
- cable spacing and burial depth;
- thermal equivalent resistances of the cable (e.g. between conductor and sheath and between sheath and armour);
- thermal resistance of the surrounding medium depending upon the soil resistivity and mutual thermal resistances between the cables;
- ambient (soil) temperature;
- allowable temperature of the dielectric.

Table 3.1 Data of EHV single-core cables (380 kV, 50 Hz)

Cable		#a	#b	#c	#d
Cross-sectional area	mm^2	1600 Cu	2500 Cu	3250 Cu	630 Cu
Conductor diameter (Milliken type)	mm	50.7	63.4	72.3	30.8
XLPE insulation diameter	mm	103.7	119.9	130.4	92.8
Metallic sheath diameter	mm	114.1 Al	130.1 Al	140.4 Al	101.8 Lead
Sheath cross-sectional area	mm^2	≅ 500	≅ 500	≅ 500	≅ 1000
PE jacket diameter	mm	125.7	141.7	152.0	117.0
Total mass	kg/m	26	37	45	28

Table 3.2 Positive sequence parameters of UGC (50 Hz) directly buried in cross-bonding at a depth 1.45 m in the soil with $\rho_{th} = 1.0\,\mathrm{K\,m/W}$; spacing $s = 0.35$ m; flat laying and soil ambient temperature $\vartheta_a = 20\,°\mathrm{C}$

Cable			#a	#b	#c	#d*
Phase cross-section		mm^2	1600 Cu	2500 Cu	3250 Cu	630 Cu
Shunt capacitance with $\varepsilon_r = 2.3$	c	µF/km	0.200	0.234	0.252	0.123
Shunt leakance with $\tan\delta = 0.0007$	g	nS/km	44	52	55	27
Phase resistance at 90 °C	r	mΩ/km	17.6	13.3	11.8	36
Series inductance	ℓ	mH/km	0.621	0.576	0.550	0.460
Characteristic impedance	$\underline{Z}_c$	Ω	$55.82\,\angle{-0.04}$ rad	$49.68\,\angle{-0.04}$ rad	$46.76\,\angle{-0.03}$ rad	$62.09\,\angle{-0.12}$ rad
Propagation factor	$\underline{k}$	1/km	$3.5 \times 10^{-3}\,\angle 1.53$ rad	$3.7 \times 10^{-3}\,\angle 1.53$ rad	$3.7 \times 10^{-3}\,\angle 1.53$ rad	$2.4 \times 10^{-3}\,\angle 1.45$ rad
Natural power at 400 kV	$\frac{400^2}{\underline{Z}_c^*}$	MVA ∠rad	$2867\,\angle 0.04$ rad	$3221\,\angle 0.04$ rad	$3422\,\angle 0.03$ rad	$2577\,\angle 0.12$ rad
Capacitive current with $U_0 = 400\,\mathrm{kV}/\sqrt{3}$	I_{cap}	A/km	14.5	17.00	18.74	8.92
Steady state ampacity	I_a	A	1498	1788	1950	650

* #d spacing = 0.12 m

The Standard IEC 60287 [3], regarding the calculation of the ampacity in different conditions, is now completely acknowledged by almost all the National Rules; numerous scientific books are devoted to UGC constructional and installation (electrical and thermal) features [4–6]. These contributions show very clearly that in transient or cyclic operation the thermal capacities (of the conductor, of the dielectric and of the surrounding soil) can temporarily increase the ampacity.

A systematic analysis of the steady state regimes compatible with the aforementioned constraints (3.1) and (3.2) can be developed by means of the transmission equations (2.13) and (2.14) of the "positive sequence model" (where $\underline{D} \equiv \underline{A}$):

$$\underline{U}_{0S} = \underline{A} \cdot \underline{U}_{0R} + \underline{B} \cdot \underline{I}_R \tag{3.4}$$

$$\underline{I}_S = \underline{C} \cdot \underline{U}_{0R} + \underline{A} \cdot \underline{I}_R \, . \tag{3.5}$$

Tables 3.1 and 3.2 report the main data for the four different EHV (380 kV AC) #a, #b, #c and #d cables. For a wide survey of the installations and a general frame on the EHV cable technology refer to Chapter 1 and [7].

In order to show the application of the procedures, examples regarding chiefly the use of cable #b will be presented.

The per unit-length capacitive susceptance of UCG (see examples of Table 3.2) is about 15–20 times that of a typical 380 kV OHL and can require (depending upon length d of the cable) strong capacitive power: the reactive susceptance affects sensibly the power flows in steady state regimes and plays furthermore a key role on the transient regimes of the line at no-load. A rough estimation brings to compute, for UGC #b, the heavy steady state capacitive power $Q_c \approx (3U_0^2 \omega c) = 11$ Mvar/km at 50 Hz and 13 Mvar/km at 60 Hz: it becomes foreseeable that shunt compensation reactors are almost always necessary to achieve good effects both in the steady state and transients regimes (see Section 3.7). It is worth noting that at $f = 60$ Hz the ampacities are lower than at $f = 50$ Hz due to the greater power losses (both Joule and dielectric ones).

It is worth remembering that the hypothesis (made in Chapter 2) of uniformly distributed shunt compensation can be adopted to easily compute the suitable transmission parameters $\underline{A}$, $\underline{B}$ and $\underline{C}$ to use in (3.4), (3.5) and in the other equations involved: in fact, it has been throughout verified that both the steady state and subtransient regimes remain almost unchanged by applying (with the same ξ_{sh}) lumped shunt compensations on length intervals not greater than 15–20 km.

3.3 First Analysis: $\underline{U}_{0S}(\delta)$, $\underline{I}_R$ Constrained

A first analysis on the possible operating conditions can be performed if in (3.4) and (3.5) both the current phasor $\underline{I}_R$ at R and the voltage phasor $\underline{U}_{0S}$ at S are constrained as follows, in accordance with (3.1) and (3.3),

$$\underline{I}_R = (I_c \angle 0) \quad \text{on the real axis,} \tag{3.6}$$

$$\underline{U}_{0S} = (U_{0c} \angle \delta) \, , \quad \delta = 0\text{–}2\pi \, , \tag{3.7}$$

so that the remaining variables $\underline{U}_{0R}$ and $\underline{I}_S$ are univocally individuated (see Section 2.5) by means of:

$$\underline{U}_{0R} = \frac{1}{\underline{A}} \cdot \underline{U}_{0S} - \frac{\underline{B}}{\underline{A}} \cdot \underline{I}_R \quad \rightarrow \quad \underline{U}_{0R} = \frac{1}{\underline{A}} \cdot (230\,\text{kV}\angle\delta) - \frac{\underline{B}}{\underline{A}} \cdot (I_c\angle 0)\,, \tag{3.8}$$

$$\underline{I}_S = \frac{\underline{C}}{\underline{A}} \cdot \underline{U}_{0S} + \frac{1}{\underline{A}} \cdot \underline{I}_R \quad \rightarrow \quad \underline{I}_S = \frac{\underline{C}}{\underline{A}} \cdot (230\,\text{kV}\angle\delta) + \frac{1}{\underline{A}} \cdot (I_c\angle 0)\,. \tag{3.9}$$

In order to develop the analysis of the $\underline{I}_S$ regimes determined by (3.9), the phasorial diagram of Figure 3.1 is useful (where only for graphical purpose, point N' has been placed sensibly far from point N): it shows, once fixed on the real axis $\underline{I}_R = I_c = 1.78\,\text{kA}$ (if cable #b is considered), how the angle δ of $\underline{U}_{0S} = (230\,\text{kV}\angle\delta)$ can range the whole interval 0–2π giving rise to two different groups:

i) any regime with $|\underline{I}_S| < I_c = |\underline{I}_R|$ compatible with the cable ampacity according to the constraint (3.2) (e.g. see the phasors OL_1, OL_2, OL_3);
ii) any regime with $|\underline{I}_S| > I_c$ not allowable (e.g. see the phasors OH_1, OH_2, OH_3).

In particular, the phasor OK_1 (when $\delta = \delta_1$) and the phasor OK_2 (when $\delta = \delta_2$) correspond to $|\underline{I}_S| = I_c = |\underline{I}_R|$, and concern two very meaningful regimes for the power flows between S and R (see Section 3.6). It is evident that for δ ranging between δ_1 and δ_2 the voltage phasors $\underline{U}_{0S}(\delta)$ determines in this first analysis all the current phasors $\underline{I}_S$ (e.g. OL_1, OL_2, OL_3) having magnitudes within the limit I_c according to (3.2).

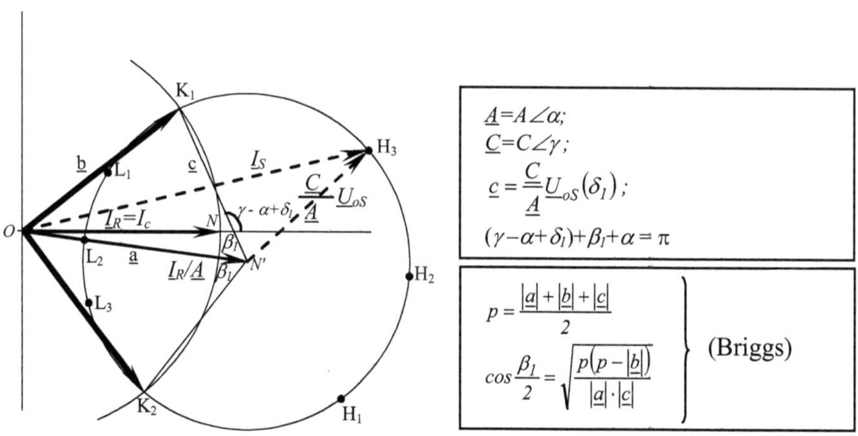

Figure 3.1 Phasorial diagram of first analysis

The values δ_1 and δ_2 can be determined by the following relations

$$\delta_1 = \pi - \gamma - \beta_1 \,, \tag{3.10}$$

$$\delta_2 = \delta_1 + 2\beta_1 \,, \tag{3.11}$$

which are evident by observing the Figure 3.1, where the necessary Brigg's formulae are also reported. It is worth including in the analysis procedures a "warning message" to display possible incompatibilities of the magnitudes $|\underline{a}|$, $|\underline{b}|$, $|\underline{c}|$ in the formation of the corresponding triangle.

The calculation of $\underline{U}_{0R}$ by means of (3.8) closes the first analysis: after this, a first set of steady state regimes compatible with the constraints (3.1)–(3.3) is completely individuated in the interval δ_1–δ_2.

For these regimes (where $|\underline{I}_R| = I_c$ and $|\underline{I}_S| \leq I_c$), it is of paramount interest to compute the complex three-phase powers at ports S and R.

3.4 Second Analysis: $\underline{U}_{0S}(\vartheta)$, $\underline{I}_S$ Constrained

A second analysis on the possible operating conditions can be performed if in (3.4) and (3.5) both the current phasor $\underline{I}_S$ and the voltage phasor $\underline{U}_{0S}$ at S are constrained as follows, in accordance with (3.2) and (3.3)

$$\underline{I}_S = (I_c\angle 0) \quad \text{on the real axis} \,, \tag{3.12}$$

$$\underline{U}_{0S} = (U_{0c}\angle\vartheta) \,, \quad \vartheta = 0\text{–}2\pi \,, \tag{3.13}$$

so that now the remaining variables $\underline{U}_{0R}$ and $\underline{I}_R$ are univocally individuated (see (2.34) in Section 2.5) by means of:

$$\underline{U}_{0R} = \underline{D}\cdot\underline{U}_{0S} - \underline{B}\cdot\underline{I}_S \quad \rightarrow \quad \underline{U}_{0R} = \underline{D}\cdot(230\,\text{kV}\angle\vartheta) - \underline{B}\cdot(I_c\angle 0) \,, \tag{3.14}$$

$$\underline{I}_R = -\underline{C}\cdot\underline{U}_{0S} + \underline{A}\cdot\underline{I}_S \quad \rightarrow \quad \underline{I}_R = -\underline{C}\cdot(230\,\text{kV}\angle\vartheta) + \underline{A}\cdot(I_c\angle 0) \,. \tag{3.15}$$

Equation 3.15 can be visualized by means of the phasorial diagram of Figure 3.2.

This diagram shows how the angle ϑ of $\underline{U}_{0S} = (U_{0c}\angle\vartheta)$, with respect to $\underline{I}_S = I_c$ on the real axis, can range the whole interval 0–2π giving rise to two different groups:

iii) any regimes with $|\underline{I}_R| < I_c$ compatible with the cable ampacity (e.g. OL', OL'', OL''');

iv) any regimes with $|\underline{I}_R| > I_c$ not allowable (e.g. OH', OH'', OH''').

In particular, the phasor OK' (when $\vartheta = \vartheta'$) and OK'' (when $\vartheta = \vartheta''$) individuate again two regimes with $|\underline{I}_R| = I_c = |\underline{I}_S|$ and with the same relations between current angles (as OK_2, OK_1 in the first analysis). It is evident that for ϑ ranging

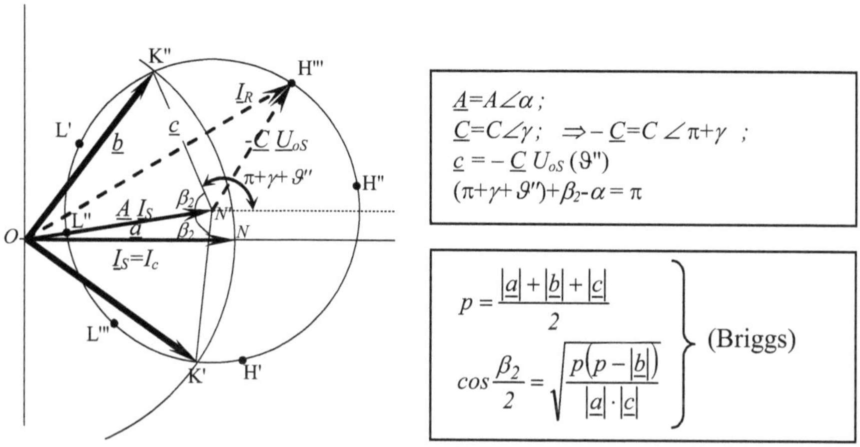

Figure 3.2 Phasorial diagram of the second analysis

between ϑ'' and ϑ' the voltage phasors $\underline{U}_{0S}(\vartheta)$ determines in this second analysis all the current phasors $\underline{I}_R$ (e.g. OL', OL'', OL''') having magnitudes within the limit I_c according to (3.1).

The values ϑ'' and ϑ' can be determined by the following relations

$$\vartheta'' = -\beta_2 - \gamma + \alpha \; , \tag{3.16}$$

$$\vartheta' = \vartheta'' + 2\beta_2 \; , \tag{3.17}$$

which are evident by observing the Figure 3.2.

The calculation of $\underline{U}_{0R}$ by means of (3.14) closes the second analysis: consequently a second set of steady state regimes compatible with the constraints (3.1)–(3.3) is completely individuated.

Also for these regimes (where $|\underline{I}_S| = I_c$ and $|\underline{I}_R| \le I_c$), it is of paramount interest to compute the complex three-phase powers at ports S and R.

3.5 Voltages and Currents Along the Cable

The analysis of the currents and voltages at sending-end S and receiving-end R does not assure "a priori" a satisfying electric behaviour along the cable: in fact, it is important to examine the currents and voltages along the cable in order to investigate possible points exceeding the fixed I_c or the voltage limits; to this aim, the use of (2.36) could be suitable.

For instance in Figures 3.3–3.6 the voltage and current behaviours are shown along the 60 km cable #b in the cases $\xi_{sh} = 0$ and $\xi_{sh} = 0.608$ (distributed compensation); the regimes of the "first analysis" are pointed out in Figure 3.1 by means of the points K_1, L_1, L_2, L_3, K_2 (equidistant between themselves on the circumference arc) and the regimes of "second analysis" are pointed out in Figure 3.2 by means of the points K'', L', L'', L''', K'.

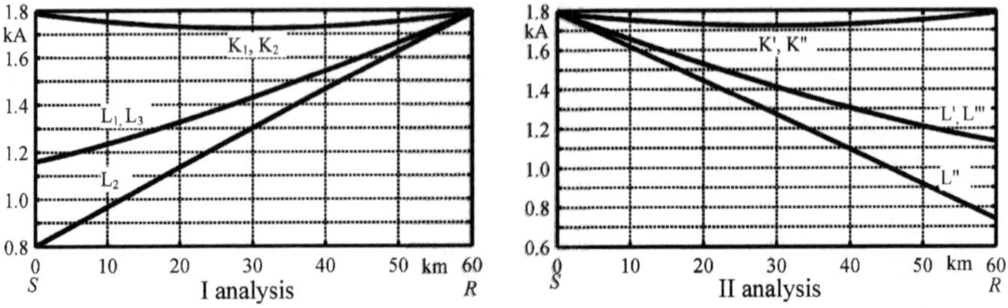

Figure 3.3 Current magnitudes along the cable #b ($\xi_{sh} = 0.0$)

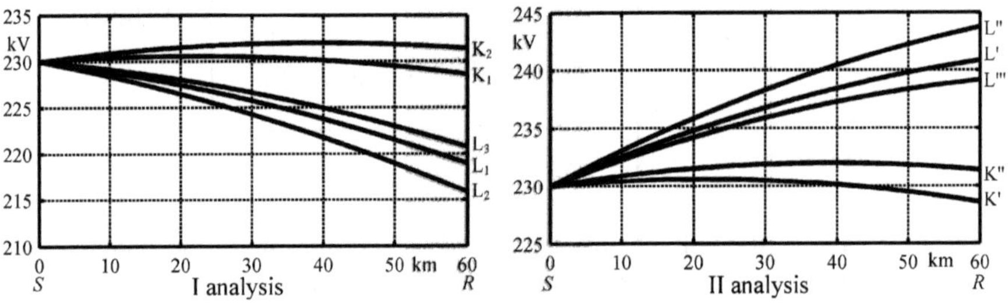

Figure 3.4 Voltage magnitudes along the cable #b ($\xi_{sh} = 0.0$)

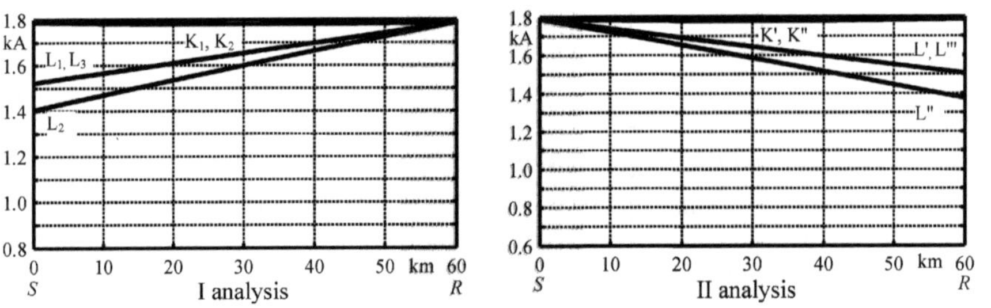

Figure 3.5 Current magnitudes along the cable #b ($\xi_{sh} = 0.608$)

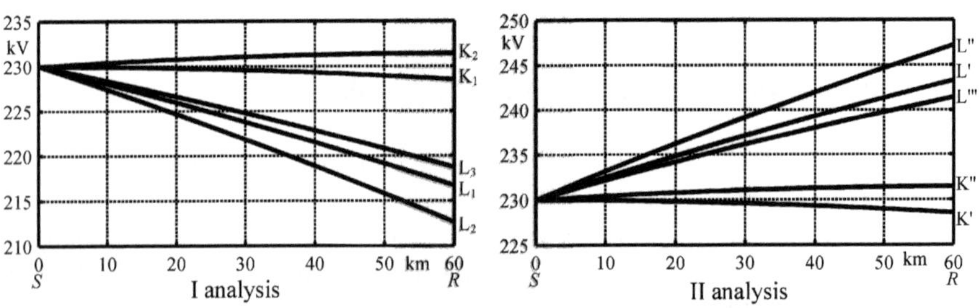

Figure 3.6 Voltage magnitudes along the cable #b ($\xi_{sh} = 0.608$)

It can be seen how the current levels are always within the ampacity limit ($\approx 1.8\,\mathrm{kA}$) and how possible voltage levels (steady state) meaningfully high can be detected at the end R.

To this aim, it is worth noting, as a typical example, the regime L'', which would give rise to (see Figure 3.6) the high level $U_{0R} \approx 247\,\text{kV}$: however it would have (see Figure 3.5) currents $|\underline{I}_S| = 1.8\,\text{kA}$ and $|\underline{I}_R| \approx 1.4\,\text{kA}$ and a corresponding (see Figure 3.31) strongly capacitive power flows ($\approx -1000\,\text{Mvar}$) of scarce interest in the network operation. In general, strong variations of current magnitude along the cable are justifiable when the quadrature components are prevailing.

A more complete and elegant method for the analysis along the cable is described in Section 3.9.2.

3.6 Power Values Compatible with Basic Constraints and with Voltage Levels at the Receiving-End

The procedures shown for the first and second analysis allow individuating all the regimes at power frequency compatible with the basic constraints (3.1)–(3.3): by using the couple of phasors $\underline{U}_{0S}$, $\underline{I}_S$ and $\underline{U}_{0R}$, $\underline{I}_R$ in each steady state regime the computation of the three-phase complex power $\underline{S}_S$ and $\underline{S}_R$ (at ports S and R) is immediate for each given line.

The corresponding graphical representation is given in the following Figures 3.8–3.11 which refer to cable #b without reactive compensation ($\xi_{\text{sh}} = 0$).

By using the criteria of first analysis, the regimes (in the interval δ_1–δ_2) compatible with the constraints $|\underline{I}_R| \equiv I_c$ and $|\underline{I}_S| \leq I_c$ have been evaluated and their receiving-end complex power have been reported forming the curve (a), whilst the corresponding sending-end complex power forming the curve (b).

Analogously, by analysing (in the interval ϑ''–ϑ') the regimes $|\underline{I}_S| \equiv I_c$ and $|\underline{I}_R| \leq I_c$, according to the second analysis, the curves (c) and (d) have been drawn; the curve (c) represents the receiving-end complex power and (d) the sending-end complex power. Therefore, once fixed $I_c = 1.78\,kA = I_a =$ *steady state cable ampacity*, the contours (a) and (c) set the boundaries of allowable "receiving-end power area" (see, for instance, grey area in Figure 3.9) whereas the contours (b) and (d) of allowable "sending-end power area" (clearly reported in the examples of Figures 3.8–3.11).

In particular, the K points, which highlight "maximum" values for P_R and P_S correspond to the homologous regimes of Figures 3.1 and 3.2. Any complex power $\underline{S}_R$ within the receiving-area gives, at ports S and R, current magnitudes $\leq I_c$; the same is valid for any $\underline{S}_S$ within the sending-area.

Since the voltage levels at node R play a key role in the network operation, it seems suitable to give them a direct visualization in the "receiving-end power area", by means of phase voltage curves parameterized with $|\underline{U}_{0R}|$ constant. It is sufficient to implement the well-known expression of receiving-end power

$$\underline{S}_R = 3 \cdot \underline{U}_{0R} \cdot \left[\frac{(\underline{U}_{0S} - \underline{A} \cdot \underline{U}_{0R})}{\underline{B}} \right]^* ; \tag{3.18}$$

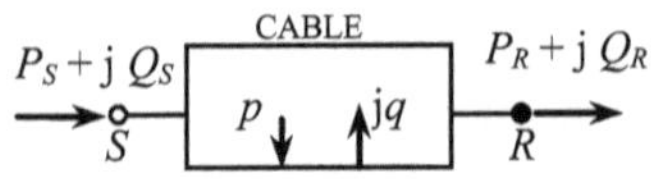

$$P_R = P_S - p \tag{3.19}$$

$$Q_R = Q_S + q \tag{3.20}$$

Figure 3.7 Power balance of a UGC transmission line

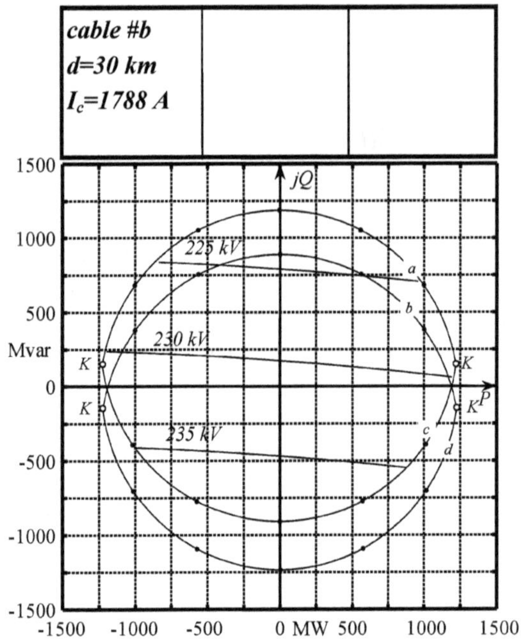

Figure 3.8 Power flows compatible with the basic constraints (without shunt compensation)

once fixed $\underline{U}_{0S} = 230\,\text{kV}\angle\sigma$ and by setting $\underline{U}_{0R}\angle\rho$ (e.g. with magnitudes 220, 225, 230, 235, 240 kV) it is possible to give suitable σ–ρ values such to determine complex power $\underline{S}_R$ in the "receiving-end area".

In such a way, it can be clearly singled out the regimes which are not acceptable owing to excessively high or low voltage levels at R (even if compatible with the basic constraints): for example, heavy reactive power flows or active ones on long runs. In particular the voltage curve intersecting the origin of the axes also points out the steady state voltage in receiving-end at no-load regime ($P_R + \text{j}\,Q_R \equiv 0$), when $|\underline{U}_{0S}| = 230\,\text{kV}$.

By the relation (3.19) of the cable power balance (see Figure 3.7) it can be observed that each receiving-end power P_R differs slightly from the corresponding sending-end power P_S, since the active power losses p are always low enough, also for UGC long runs.

Equation 3.20 shows that each receiving-end reactive power Q_R results clearly higher than the corresponding Q_S, since the uncompensated cable itself "generates"

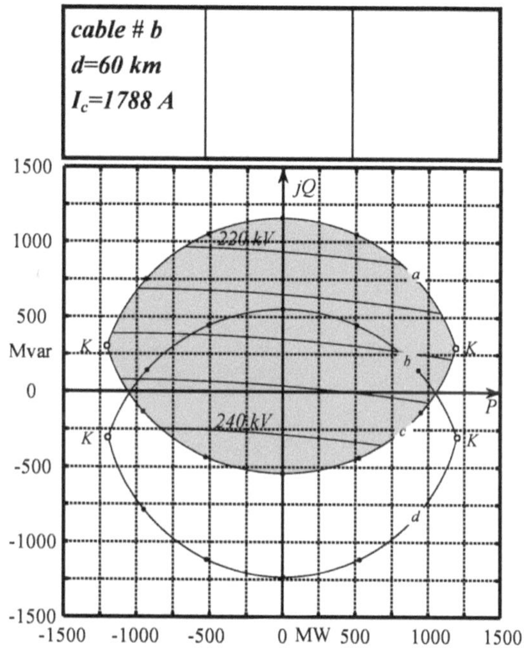

Figure 3.9 Power flows compatible with the basic constraints (without shunt compensation)

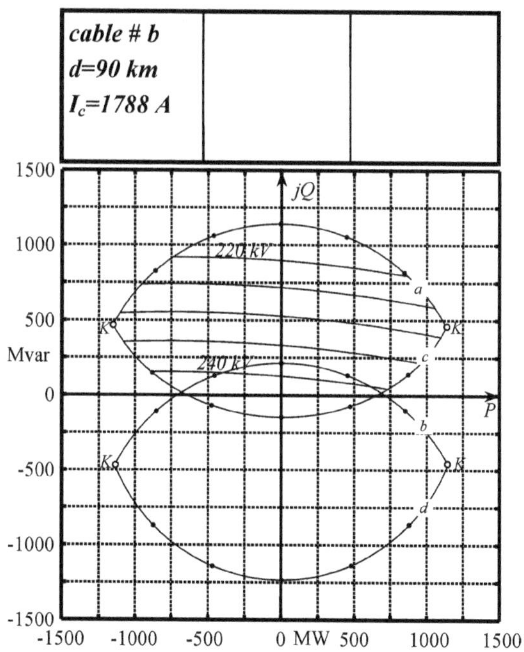

Figure 3.10 Power flows compatible with the basic constraints (without shunt compensation)

Figure 3.11 Power flows compatible with the basic constraints (without shunt compensation)

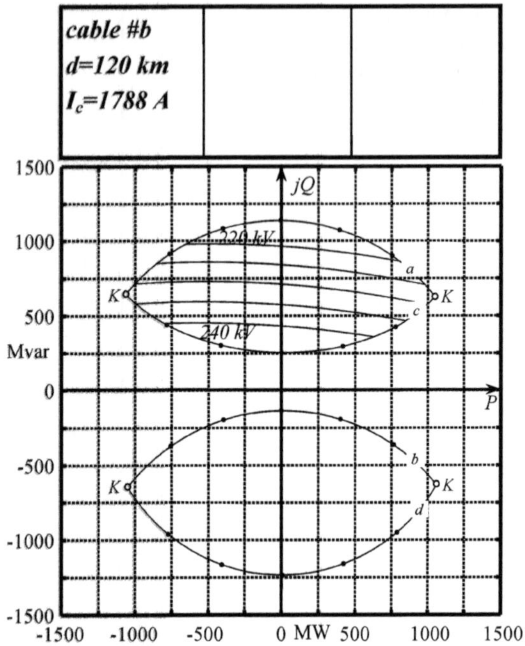

high inductive power q, increasing with the length d. That helps to explain how, by increasing the length d, the area contours have progressive translations along the imaginary axis of the complex plane. By considering the cases from $d = 30\,\text{km}$ to 120 km (without compensation), it can be ascertained that the receiving and sending areas are progressively "moving away" from the x-axis (beyond a perceptible "shrinkage" along the y-axis): this highlights that in the power flows a higher and higher reactive power (rather than an active one) must be accepted, so opposing the chief aim of the EHV transmission links.

By this important point of view, the length of 25–30 km (see Figure 3.8) can be considered as a limit for the cable #b unless the shunt reactive compensation is performed with a given degree (see Section 3.7).

It is worth adding that, beyond the aforementioned drawbacks, in order to have suitable power flows, other heavy problems, regarding the system security itself, are the regimes (steady state and transient) of no-load cable energization and de-energization besides the load-rejection, which almost always force the use (see Section 3.7) of a suitable shunt compensation and give the installation its final arrangement.

3.7 No-Load Energization and De-Energization

It is worth remembering that in the network operation, very frequently recurring events of greatest concern are the energization (and de-energization) of a no-load line since it is always necessary to prepare the operating structure of the grid.

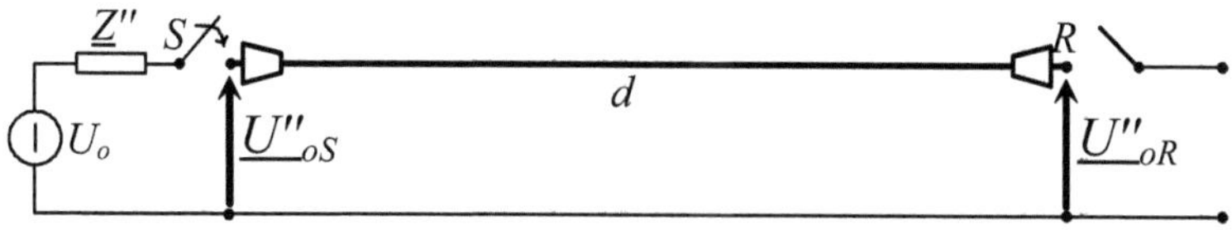

Figure 3.12 No-load energization at port S

The sketch of Figure 3.12 shows the switch-on of the circuit breaker in S in order to energize the cable and serves as a first, not complete but meaningful, approach devoted to highlight the limit conditions for line energization at no-load. In a generic network, the power supply in S can be modelled as an equivalent generator which is characterized by its electromotive force U_0 (supposed to be equal to 230 kV) and the short-circuit subtransient impedance $\underline{Z}''$ (for simplicity purely inductive $\underline{Z}'' = \mathrm{j}\,X''$).

The magnitude of no-load power frequency sub-transient voltage $\underline{U}''_{0R}$ at R, due to the closing at S, constitutes a reference ("indicial" value) to foresee the peak value of switching surges due to subsequent evolution of the phenomena and is completely defined through the following formulae:

$$\underline{U}''_{0S} = \left(\frac{U_0}{\underline{Z}''\cdot + \frac{\underline{A}}{\underline{C}}}\right)\cdot\frac{\underline{A}}{\underline{C}}, \tag{3.21}$$

$$\underline{U}''_{0R} = \frac{\underline{U}''_{0S}}{\underline{A}}, \tag{3.22}$$

$$\underline{U}''_{0R} = \frac{U_0}{\underline{A} + \underline{Z}''\cdot\underline{C}}, \tag{3.23}$$

with $\underline{A}/\underline{C}$ the impedance (almost completely capacitive) as seen from S with R at no-load, and expressing with (3.22) the Ferranti's effect.

With regard to $\underline{Z}'' = \mathrm{j}\,X''$ evaluation, it is possible to refer to the subtransient impedance U_0/I''_{sc} (from network studies): since the values of subtransient short circuit current I''_{sc} (three-phase at S) in EHV networks can be foreseen in the range 10–50 kA, X'' corresponds to 23–4.6 Ω.

In order to respect the standard switching levels (e.g. 1050 kV) with a conservative margin, it seems advisable that the phasor $\underline{U}''_{0R}$ does not exceed the magnitude $U_{\mathrm{m}}/\sqrt{3} = 242.5$ kV: it is easy to ascertain that such target can be reached introducing by attempts in (3.23) a suitable compensation degree which modifies suitably the parameters $\underline{A}$ and $\underline{C}$.

In the hypothesis that, extinguished the transient phenomena, the voltage regulation restores again at port S the rated value $U_0 = 230$ kV, it is possible to compute

$$\underline{I}_{\mathrm{NL}} = \frac{U_0}{(\underline{A}/\underline{C})} \tag{3.24}$$

which expresses the no-load steady state (almost entirely capacitive) current $\underline{I}_{\mathrm{NL}}$ (which the circuit breaker b must interrupt in case of de-energization of the no-load cable): for this current the Standards on the Circuit Breaker (Section 4.107 of [8]) suggest the limit value of 400 A.

After all, both the following additional constraints must be fulfilled (by assigning by attempts a suitable compensation degree ξ_{sh})

$$|\underline{U}''_{0R}| \leq 242.5\,\text{kV}\,, \tag{3.25}$$

$$|\underline{I}_{\text{NL}}| \leq 400\,\text{A}\,. \tag{3.26}$$

The curves of Figures 3.13 and 3.14 show that the constraint (3.26) is almost always decisive for the determination of ξ_{sh}, unless there is an agreement with the manufacturer for a circuit breaker with higher I_{NL}.

Figure 3.15 clearly defines the values of shunt reactive compensation degree ξ_{sh} in order to fulfil (3.25) or (3.26) showing which one is the more limiting criterion as a function of the line length.

It appears almost trivial to note that analogous considerations in order to verify the additional constraints must be made also for the energization by the port R (besides by the port S), by introducing for the parameters U_0 and X'' suitable values in consideration of the supply grid linked to R: the detection and the choice of the "best-end switching" [9] are mentioned by CEGB as effective experienced practices in network operations.

Moreover, the cases where the steady state capacitive power absorbed by the cables ($Q_{\text{NL}} = 3U_0 I_{\text{NL}}$) exceeds the ability of synchronous generators (dangerous self-excitation conditions), located in close proximity of the cable installation, must be avoided [10–12].

In any case it is absolutely advisable that the Transmission System Operator (TSO), when planning a new link, performs both detailed power flows and network simulations.

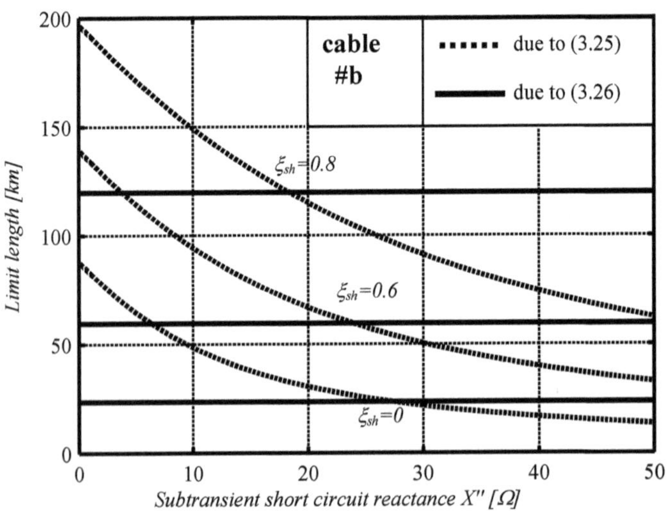

Figure 3.13 Limit lengths due to the constraints (3.25) and (3.26) for cable #b

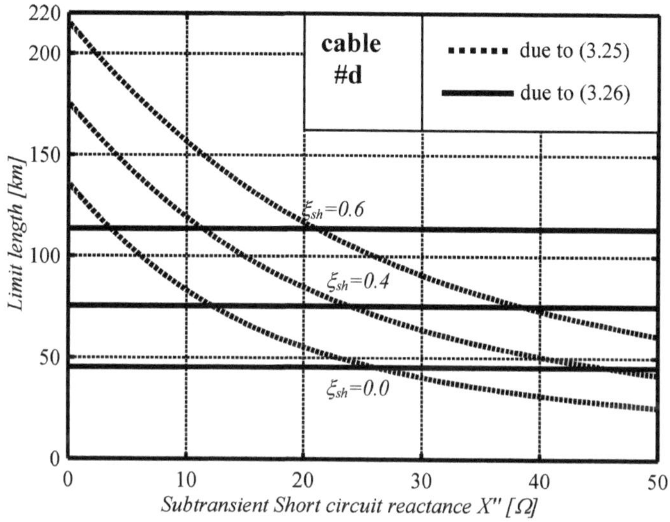

Figure 3.14 Limit lengths due to the constraints (3.25) and (3.26) for cable #d

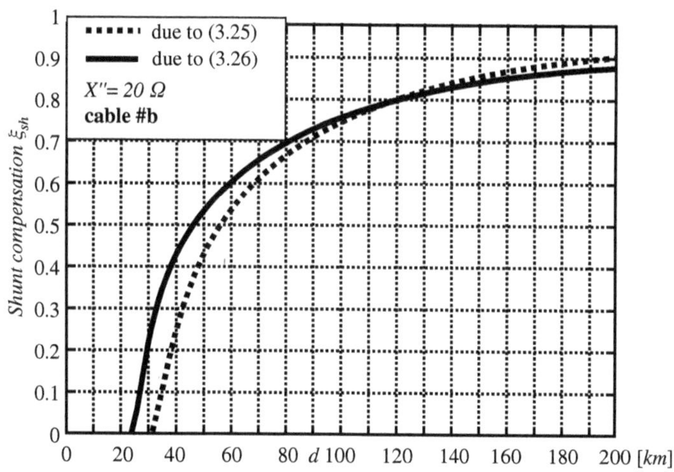

Figure 3.15 Reactive compensation degree as a function of cable length due to the constraints (3.25) and (3.26) for cable #b and $X'' = 20\,\Omega$

3.8 Power Capability Charts

Once the degree of shunt reactive compensation has been chosen (in accordance with Section 3.7) in order to fulfil the additional constraints (3.25) and (3.26), it is possible to draw the final capability charts compatible with the basic constraints, showing the voltage levels U_R at R for a good line operation, reporting above a summing-up table with the values of the following parameters which fully characterize the installation and with some noteworthy regimes:

d cable length;
I_c chosen current $\leq I_a$;
X'' 50 Hz subtransient reactance (seen at port S);
ξ_{sh} shunt compensation degree according to Section 3.7 (unif. distributed one);
U''_{0R} RMS value of the subtransient voltage in the no-load cable energization from the port S, with a given degree ξ_{sh};
I_{NL} RMS value of the current absorbed in S by the cable at no-load in R, in the steady state regime with $|\underline{U}_{0S}| = 230$ kV and with the given degree ξ_{sh};
Q_{NL} Capacitive reactive power absorbed in S by the cable at no-load in R in the steady state regime with $|\underline{U}_{0S}| = 230$ kV and with the given ξ_{sh};
P_{R1} receiving-power with $\cos\varphi = 1$ *(indicated with □ in the figures 3.16–3.27)*;
ΔP_1 active power losses at P_{R1};
ΔQ_1 whole reactive power required by the cable at P_{R1}.

The so-depicted "capability charts" (see the following examples) are an extremely useful tool to quantify and compare, in a visually immediate way, different planning solutions.

The capability charts of Figure 3.16 show how for the cable #b at 50 Hz the shunt reactive compensation can be avoided only for lengths inferior to 23.6 km, if the capacitive no-load current I_{NL} must not exceed 400 A (with $U_{0S} = 230$ kV); Figure 3.18 highlights (in the table above) that with $d = 30$ km the condition $\xi_{sh} = 0$ would maintain $|U''_{0R}|$[1] within satisfying limits, but would give rise to $I_{NL} = 509$ A.

Figure 3.19 shows how the choice of $\xi_{sh} = 0.214$ can give the link the final settlement: it is of note (beyond the limitation of I_{NL} to the suitable value 400 A) also a slight increasing of P_{R1} and a good subtransient voltage level, which leaves to foresee for the allowable limit subtransient reactance values up to 27 Ω, in accordance with (3.23) and (3.25). Figure 3.17 shows the capability charts of the cable #b at 60 Hz without shunt compensation: it is worth noting the worsening of the limit length ($d = 19.7$ km) and of the ampacity. In general, the frequency (60 vs 50 Hz) plays its important role in the UGC analyses.

If other lengths of the cable (see examples with 60 km in Figures 3.20 and 3.21; 90 km; 120 km) are considered, it can be observed, with immediate visual comparisons, that the high benefits given by the shunt reactive compensation (after Section 3.7) involve not only the energization – de-energization phenomena, but also the power flows.

It is worth underlining the case with $d = 90$ km, (Figures 3.22 and 3.23) where the compensation degree $\xi_{sh} = 0.74$ (which gives additional costs) modifies drastically the no-load phenomena U''_{0R} and I_{NL}, and requalifies dramatically the allowable power flows: it can be noted the power P_{R1} increase from 652.5 MW to

[1] In all the figures, for the sake of brevity, $|\underline{U}''_{0R}|$ will be written as U''_{0R}.

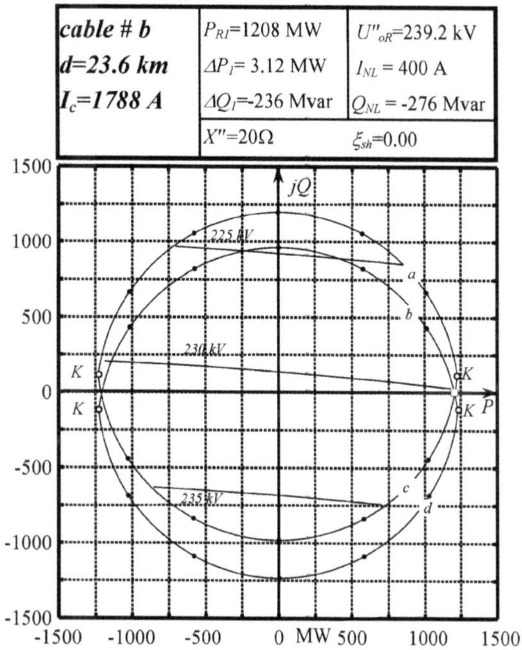

Figure 3.16 Capability charts without shunt compensation

1220 MW (87%), the reduction of ΔQ_1 and, in general, improvements of the power flows in the four quadrants.

In the case of $d = 120$ km (always cable #b) the comparison of Figures 3.25 and 3.24 appears very impressive at first sight, because there is an almost complete "overlapping" (beyond a sensible widening and a good symmetrization in the complex plane) of the "receiving" and "sending" areas, by means of the very high $\xi_{sh} = 0.81$, for which (differently from the foregoing cases) the constraint (3.25) is more influencing rather than (3.26). The shunt compensation degree $\xi_{sh} = 0.810$ is mandatory for having the subtransient voltage at R from the value 313 kV (with $\xi_{sh} = 0$) to the value 242.5 kV after (3.25) and is also sufficient to carry the no-load current to the allowable value 390 A. For the computation of the capital costs, it is worth remembering that such a compensation implies the installation of shunt reactors in at least five sections along the line for a total rated power of roughly $0.81 \cdot 11 \cdot 120 \approx 1070$ Mvar. The capability chart of Figure 3.24 (with $\xi_{sh} = 0$) is rather interesting for the steady state regimes, since both the power areas, which are sensibly far from the real axis, show power flows with strongly reactive components.

It is also of note that receiving area (with $\xi_{sh} = 0$) does not include the axes origin, so showing that the regime $P_R + \mathrm{j}\, Q_R = 0$ (or steady state regime at no load at R) implies the exceeding in the cable of the set current limit, as clearly shown in Figure 3.24 from $I_{NL} = 2170\,\mathrm{A} \gg 1788\,\mathrm{A} = I_c$.

The analyses up to here have been performed for cable #b over different lengths: if cable #d is considered, the features become slightly worse. For instance, the com-

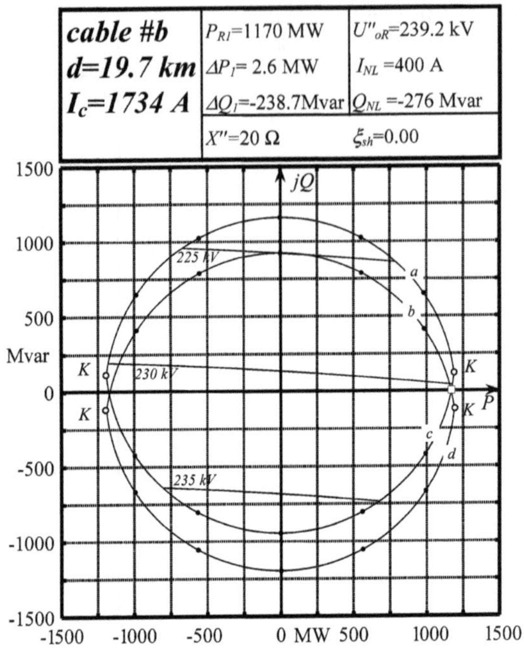

Figure 3.17 Capability charts without shunt compensation at $f = 60\,\text{Hz}$

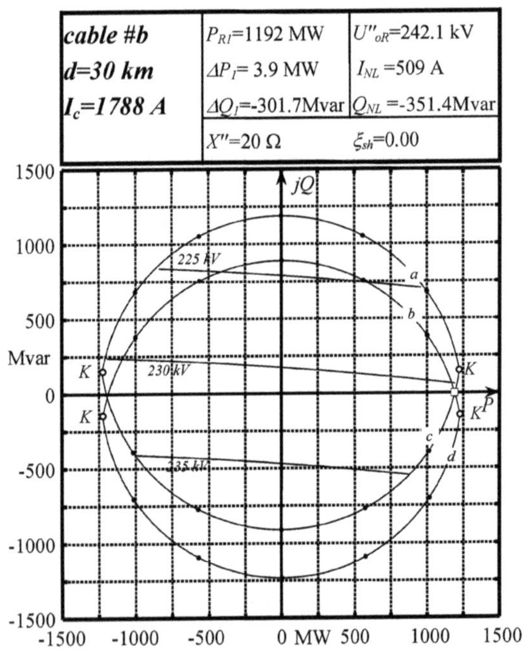

Figure 3.18 Capability charts without shunt compensation

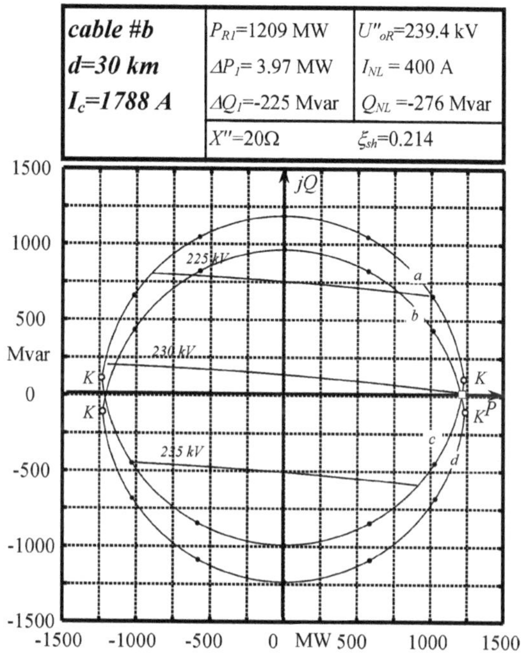

Figure 3.19 Capability charts with shunt reactive compensation

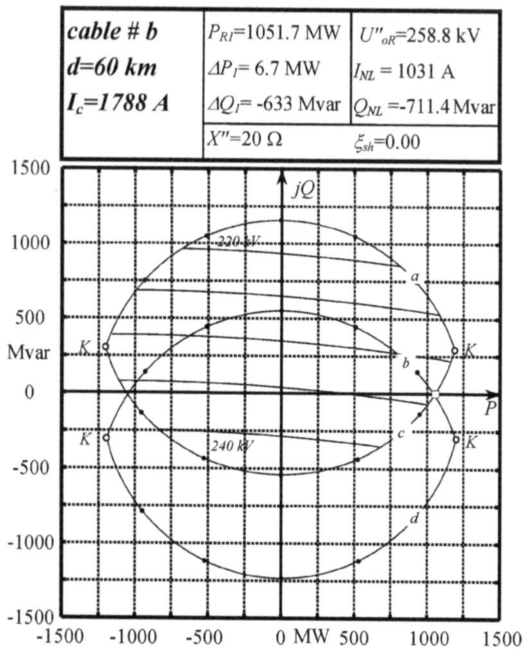

Figure 3.20 Capability charts without shunt reactive compensation

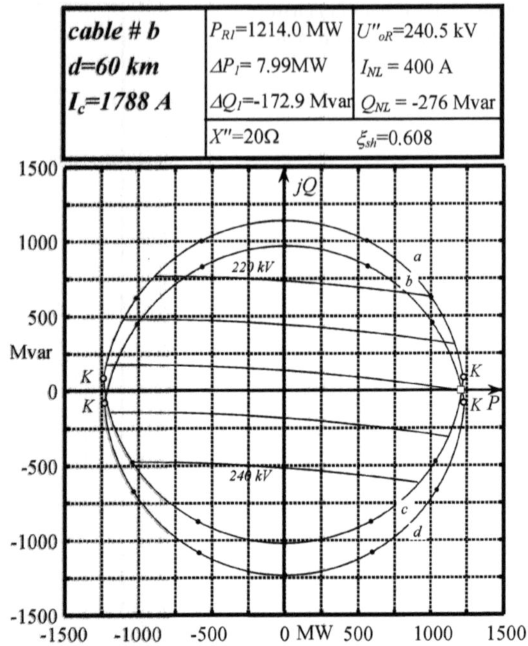

Figure 3.21 Capability charts with shunt reactive compensation

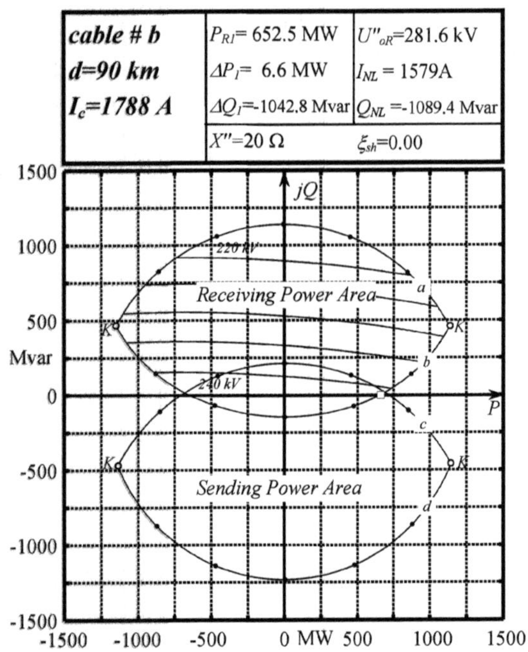

Figure 3.22 Capability charts without compensation

Figure 3.23 Capability charts with shunt reactive compensation

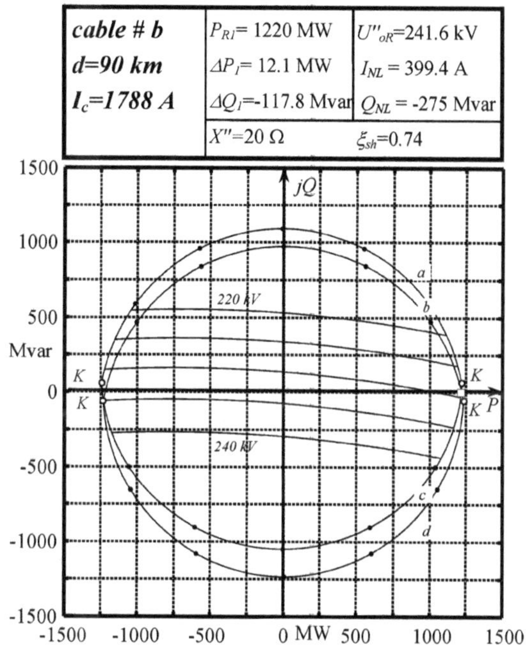

Figure 3.24 Capability charts without shunt compensation

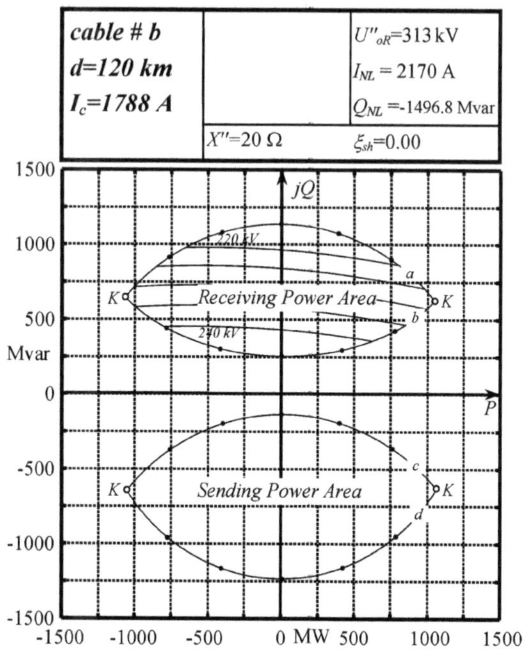

parison in Figure 3.26, performed for $d = 60\,\text{km}$, shows that cable #d suffers in the maximum transmissible active power a stronger reduction than in the ampacity and a worsening in the corresponding power factor.

Figure 3.27 shows the capability chart with a compensation degree $\xi_{\text{sh}} = 0.254$.

3.8.1 *Theoretical Limits of the Length d*

The progressive "shrinkages" and the "shifting" in the complex plane of the sending and receiving areas, which can be easily detected in the cases with $\xi_{\text{sh}} = 0$, leave to guess that (for a given cable) there is a limit length beyond which the "basic constraints" cannot be satisfied: for example the Figure 3.28 shows the shrinkages and the locations with lengths 100, 150, 180, 190, 200 km.

For a precise determination, Figures 3.1 and 3.2 are very useful: it can be verified that the points of the regimes i) and iii) vanish if

$$\left(\frac{1+|\underline{A}|}{|\underline{C}|}\right) < \frac{|\underline{U}_{0S}|}{I_{\text{c}}}\ ;$$

for instance, the limit length of 201.87 km can be computed for cable #b, once set the values $\underline{U}_{0S} = 230\,\text{kV}$ and $I_{\text{c}} = 1788\,\text{A}$.

For this length, the sending and the receiving areas degenerate in the points S_L and R_L respectively which highlight a steady state regime interesting only from a speculative standpoint and deserve a brief comment: from the coordinates of the points S_L and R_L it is possible to infer that each port S and R introduces symmetrically in the line an almost capacitive power equal to that required from the cable of half length at no load.

The analysis along the cable (by means of the algorithms shown in Section 3.9.2) shows that at S there is an almost capacitive current equal to the ampacity 1788 A; at half the line there is an almost null current; again at R there is a current equal to 1788 A.

3.9 Steady State Regimes Within Power Areas

As seen in the previous paragraphs, the capability charts are extremely useful means in order to quantify and compare accurately the cable possible performance for different types and lengths.

It is anyway important to describe some additional procedures (see Sections 3.9.1 and 3.9.2) that allow visualizing in detail the possible regimes within the receiving area.

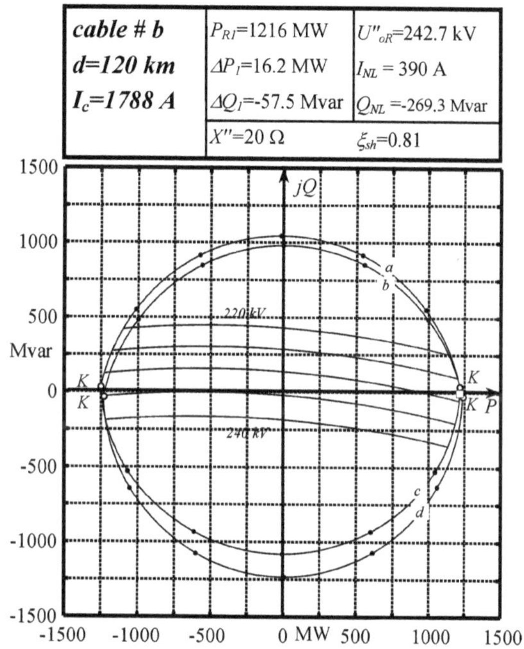

Figure 3.25 Capability charts with shunt reactive compensation

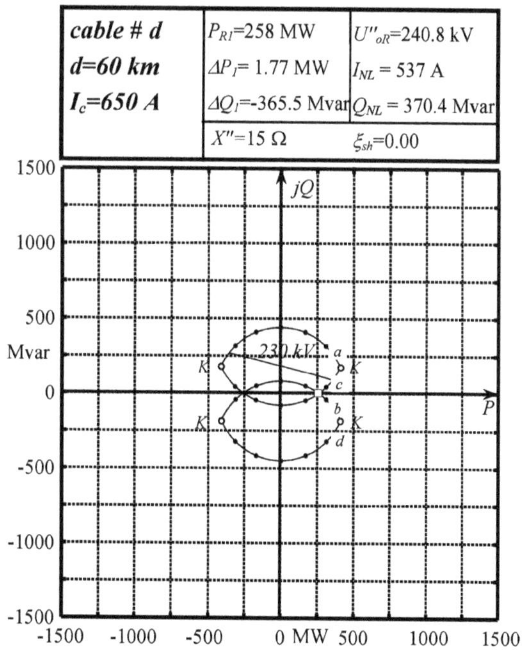

Figure 3.26 Capability charts of cable #d without shunt compensation

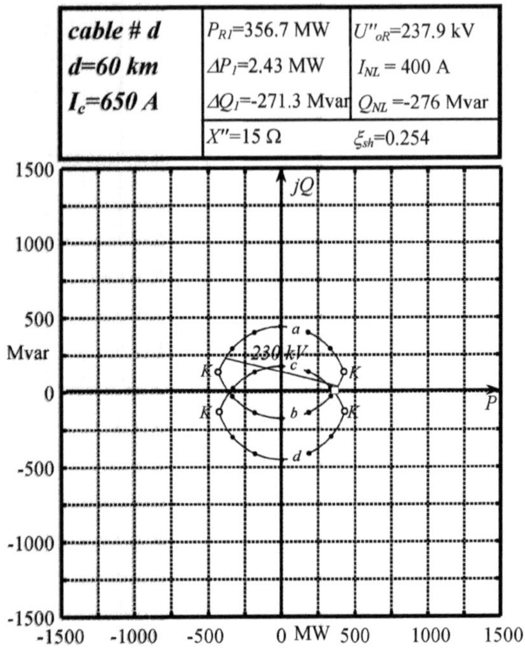

Figure 3.27 Capability charts of cable #d with shunt compensation

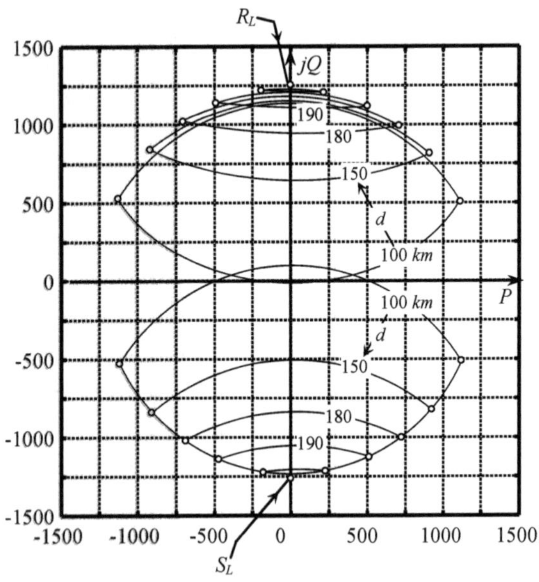

Figure 3.28 Worsening of the capability charts of cable #b by increasing the line length *d* (km), without shunt reactive compensation

3.9.1 Enhanced Capability Charts

It could be useful for transmission system operators to enhance the capability charts (see for example Figure 3.29 for the receiving-end area in the case of cable #b with

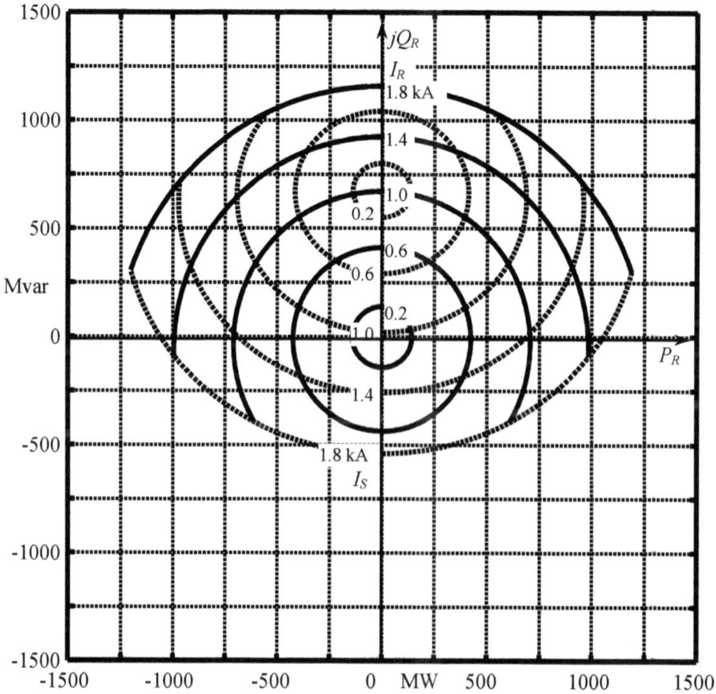

Figure 3.29 Enhanced capability charts at the receiving-end for cable #b; $d = 60\,\text{km}$; $\xi_{\text{sh}} = 0$

a length of 60 km and with $\xi_{\text{sh}} = 0$) adding other curves referring to some current values I_c smaller than cable ampacity, so that for each transmissible complex power chosen within the area the margins are immediately evident.

For the computation of these curves the methods of the first and second analysis can be suitably used by fixing on the real axis current phasors $(\underline{I}^{\bullet}_R \angle 0)$ and $(\underline{I}^{\bullet}_S \angle 0)$ with magnitudes smaller than the limit I_c.

It is sufficient to specify that in the first analysis δ is not constrained when

$$\left|\frac{\underline{I}^{\bullet}_R}{\underline{A}}\right| + \left|\frac{\underline{C}}{\underline{A}}\underline{U}_{0S}\right| < I_c\,, \tag{3.27}$$

and in the second analysis ϑ when

$$|\underline{A}\,\underline{I}^{\bullet}_S| + |\underline{C}\,\underline{U}_{0S}| < I_c\,. \tag{3.28}$$

Another procedure can be based on (3.15) inferable by (2.34) which has been premised in Section 3.4 and here written again:

$$\underline{I}_R = -\underline{C}\,\underline{U}_{0S} + \underline{A}\,\underline{I}_S\,. \tag{3.15}$$

Equation 3.15 (valid for a complete analysis) can be interpreted by the construction of the two triangles (ON″N and ON′N) of Figure 3.30, which shows how, once

Figure 3.30 Triangles for Briggs formulae application

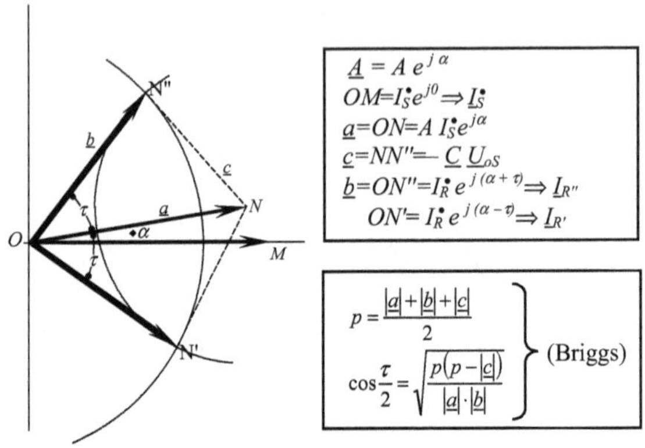

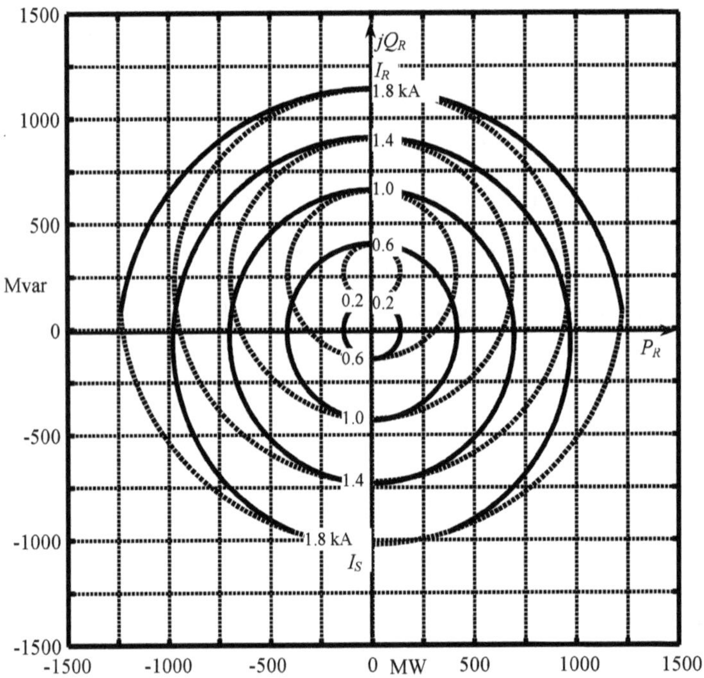

Figure 3.31 Enhanced capability charts at the receiving-end for cable #b; $d = 60\,\text{km}$; $\xi = 0.608$

fixed for the magnitudes in S and R a pertaining couple of values $\underline{I}_S^{\bullet}$, $\underline{I}_R^{\bullet}$ lower than I_c (and holding $|\underline{U}_{0S}| = 230\,\text{kV}$) it is possible (respecting the geometrical compatibility of the triangle sides and by using the Briggs' formulae) to individuate the angles $\alpha - \tau$ and $\alpha + \tau$ and hence to know the two possible regimes (the former composed of the phasors $\underline{I}_S^{\bullet}$ and $\underline{I}_{R'}$, the latter of $\underline{I}_S^{\bullet}$ and $\underline{I}_{R''}$): once obtained, with a twofold application of (2.38), the phasors $\underline{U}_{0R'}$ and $\underline{U}_{0R''}$ in the two different regimes, it is simple to calculate the two corresponding complex powers at R.

The capability chart of Figure 3.29 shows at a glance the two steady state regimes at no load (as seen at port S, since $|\underline{I}_S| \cong 1\,\text{kA}$ and $\underline{I}_R = 0$ or alternatively at the port R, since $|\underline{I}_R| \cong 1\,\text{kA}$ and $\underline{I}_S = 0$). In order to have no-load currents $\leq 400\,\text{A}$, it is necessary to adopt a shunt compensation $\xi_{\text{sh}} = 0.608$ as already shown in Figure 3.21; in this case, the enhanced capability chart is represented by Figure 3.31.

3.9.2 Application of Ossanna's Method

Starting from the single-phase model of a three-phase network seen at port R, the Ossanna's method [13] gives a simple (but very powerful and elegant) algorithm to determinate at the port itself the possible voltage regimes (and consequently the current ones as well) due to a fixed complex power (for each phase) $\underline{S}_{0R} = S_{0R}e^{j\varphi}$: the analytical direct computation avoids the iterative procedures up till now largely used. It constitutes a powerful tool to further investigate the steady state regimes inside and outside the receiving areas of the capability charts and is able furthermore to determine the limit conditions of physical feasibility.

Firstly, it appears clear that the cable line system can be studied by means of the typical scheme as seen by the port R of Figure 3.32, by considering that in such case

$$\underline{U}_{0\text{RNL}} = \frac{U_{0S}}{\underline{A}}\,, \tag{3.29}$$

$$\underline{Z} = \frac{\underline{U}_{0\text{RNL}}}{\underline{I}_{R\text{sc}}} = \frac{U_{0S}/\underline{A}}{U_{0S}/\underline{B}} = \frac{\underline{B}}{\underline{A}} = Ze^{j\psi}\,, \tag{3.30}$$

being $I_{R\text{sc}}$ = short circuit current at R so that (3.31) is valid

$$\frac{\underline{U}_{0\text{RNL}} - \underline{U}_{0R}}{\underline{Z}} = \underline{I}_R = \left(\frac{\underline{S}_{0R}}{\underline{U}_{0R}}\right)^{*}. \tag{3.31}$$

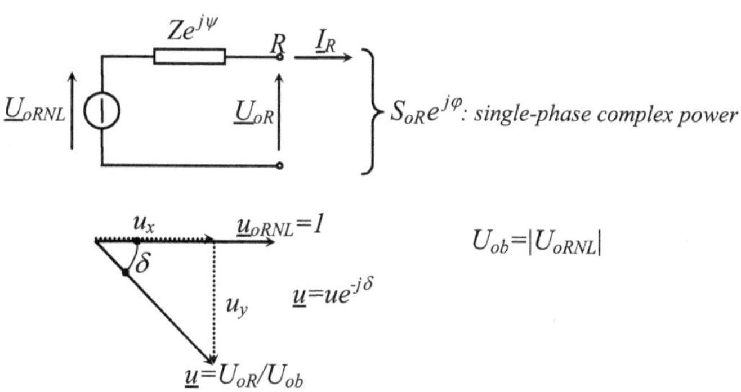

Figure 3.32 Network model and phasors $\underline{u}$ (per unit)

If the "per unit" method is adopted (exactly as J. Ossanna made at the beginning of the past century), and by assuming as basis quantities

$$U_{0b} = |\underline{U}_{0\mathrm{RNL}}| \,, \quad S_{0b} = \frac{U_{0b}^2}{|\underline{Z}|} \,, \quad Z_b = |\underline{Z}| \,,$$

and by posing $\underline{U}_{0RNL}$ on the real axis, it is possible to obtain the following p.u. relations:

$$\frac{1-\underline{u}}{\mathrm{e}^{\mathrm{j}\psi}} = \underline{i} = \frac{\underline{s}^*}{\underline{u}^*} \quad \Rightarrow \quad \underline{u}^* - u^2 = \underline{s}^* \mathrm{e}^{\mathrm{j}\psi} \quad \Rightarrow \quad u\mathrm{e}^{\mathrm{j}\delta} - u^2 = s\mathrm{e}^{\mathrm{j}(\psi-\varphi)}$$

$$\Rightarrow \quad u^2 - u\mathrm{e}^{\mathrm{j}\delta} + s\mathrm{e}^{\mathrm{j}\chi} = 0 \,, \tag{3.32}$$

where

$$\frac{\underline{U}_{0RNL}}{U_{0b}} = \underline{u}_{0\mathrm{RNL}} = 1 \,; \quad \underline{u} = \frac{\underline{U}_{0R}}{U_{0b}} \,; \quad u = |\underline{u}| \,; \qquad \underline{u} = u\mathrm{e}^{-\mathrm{j}\delta}$$

(see phasor diagram) ;

$$\frac{\underline{Z}}{Z_b} = \mathrm{e}^{\mathrm{j}\psi} \,; \qquad \underline{s} = \frac{\underline{S}_{0R}}{S_{0b}} \,; \quad \underline{s} = s\mathrm{e}^{\mathrm{j}\varphi} \,; \quad \psi - \varphi = \chi \,.$$

By explicating in (3.32) the real and the imaginary parts, a system of equations is reached

$$\begin{cases} u_x^2 + u_y^2 - u_x + s\cos\chi = 0 \,, & (3.33) \\ -u_y + s\sin\chi = 0 \,, & (3.34) \end{cases}$$

so that (3.34) gives immediately

$$u_y = s \cdot \sin\chi = \eta_{\mathrm{Oss}} \quad (\eta_{\mathrm{Oss}} = \text{Ossanna's parameter}) \tag{3.35}$$

and allows transforming (3.33) in

$$u_x^2 + (s \cdot \sin\chi)^2 - u_x + s \cdot \cos\chi = 0$$

$$\Rightarrow \quad u_x^2 - u_x + [s \cdot \cos\chi + (s \cdot \sin\chi)^2] = 0 \,. \tag{3.36}$$

From (3.36) it yields:

$$u_x = \frac{1 \pm \sqrt{1 - 4[s\cos\chi + (s\sin\chi)^2]}}{2} = \frac{1}{2} \pm \sqrt{\frac{1}{4} - s\cos\chi - (s\sin\chi)^2} \,; \tag{3.37}$$

$$u_x = 1/2 \pm \zeta_{\mathrm{Oss}} \,; \tag{3.38}$$

$$\zeta_{\mathrm{Oss}} = \sqrt{\frac{1}{4} - s\cos\chi - (s\sin\chi)^2} \quad (\zeta_{\mathrm{Oss}} = \text{Ossanna's parameter}) \,. \tag{3.39}$$

It is also to highlight that the necessary location of ζ_{Oss} in the real part gives the conditions of physical feasibility (described in [13]).

So it is possible to determine *analytically* the two phasors $\underline{u}_1$ and $\underline{u}_2$ (p.u.) possible at the port R, each of them is compatible with the same fixed power $\underline{s}$ (p.u.):

$$\underline{u}_1 = [(1/2) + \zeta_{Oss}] - \mathrm{j}\,[\eta_{Oss}] , \quad (3.40)$$

$$\underline{u}_2 = [(1/2) - \zeta_{Oss}] - \mathrm{j}\,[\eta_{Oss}] , \quad (3.41)$$

$$\zeta_{Oss} \geq 0 , \quad (3.42)$$

being that (3.42) is the constraint of physical feasibility so that it is advisable to foresee a warning message for ζ_{Oss} becoming a complex number. The absolute quantities are given by $\underline{U}_1 = u_1\,U_{0b}$ and $\underline{U}_2 = u_2\,U_{0b}$.

These relations have constituted a fundamental premise for the wide and interesting investigations developed by Ossanna [13, 14] on the performance of AC overhead lines foreseeable in those years, but now they give this book a meaningful contribution in order to enhance the analysis of capability charts.

In our case, it can be ascertained that the phasors $\underline{u}_2$, by (3.41), present always very small magnitudes (with consequently highest currents) and give (beyond the obvious theoretical interest) a precise reference chiefly for the possible investigations of voltage collapse: therefore, in the capability chart study (devoted to allowable regimes) only (3.40) must be considered.

From (3.40) and (3.41) it is immediate to derive [13] the elegant relation $\underline{u}_1 + (\underline{u}_2)^* = 1 = \underline{u}_2 + (\underline{u}_1)^*$ which clearly highlights how if $\underline{u}_1$ is close to 1 the same cannot be for $\underline{u}_2$.

Therefore it becomes easy to implement few and simple files on the basis of (3.40) in order to automatically illustrate in detail the steady state regimes which exist not only at R and S ports, but also (by implementing the obvious matrix techniques described in Section 2.5) in other sections along the line once any complex power $\underline{S}_R = 3\underline{S}_{0R}$ is set: in this way an exhaustive investigation can be performed not only inside but also outside the receiving area, so that it will be also possible to individuate the physical feasibility independently from the constraints (3.1) and (3.2).

For example (in case of cable #b, $d = 60$ km, $\xi_{sh} = 0.608$), as a completion of the analysis along the cable, with Ossanna's method, it is possible to obtain, by

Table 3.3 Three-phase receiving power $\underline{S}_R = 3\underline{S}_{0R}$

# Regime	Three-phase complex power $\underline{S}_R$ MW + j Mvar
1	1000 + j 0
2	0 + j 0
3	−1000 + j 0
4	500 + j 500
5	0 + j 500
6	−500 + j 500
7	750 − j 250
8	0 − j 250
9	−750 − j 250

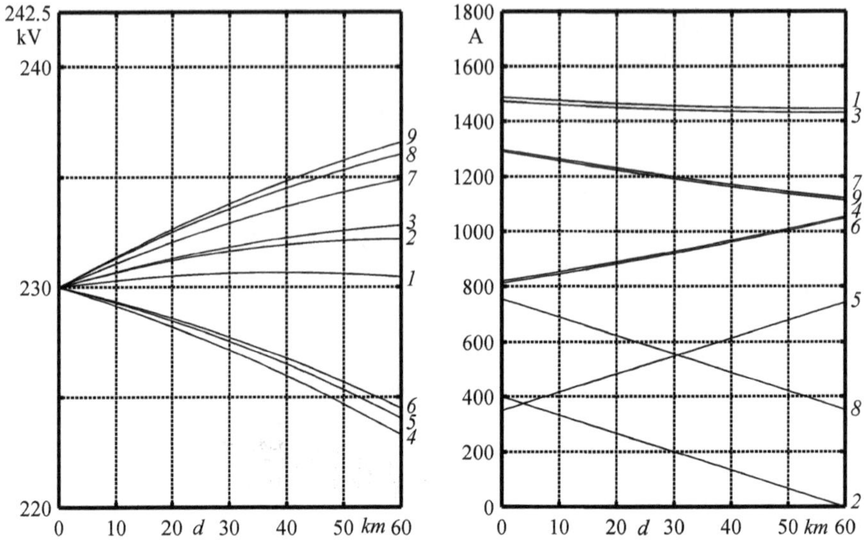

Figure 3.33 Voltage and current magnitudes along the cable #b, $d = 60\,\text{km}$ ($\xi_{sh} = 0.608$), in the regimes 1–9 of Table 3.3

hypothesizing the three-phase power in Table 3.3, the curves (along the cable) of Figure 3.33.

It is trivial to observe that, by using the suitable references, the fixed power can be located in any quadrant.

The possibility of knowing the magnitudes of the currents at different locations of the cable line becomes extremely useful in order to refine some temperature checks in those parts of the route having high thermal resistivity.

3.10 Cables with Gas Insulation (GILs)

The proposed procedure has been suitably applied to another kind of cable with gas mixture insulation namely Gas Insulated Transmission Line: Tables 3.4 and 3.5 report the constructive data of a typical 380 kV GIL and the positive sequence parameters, respectively.

As already mentioned in Chapter 2, the enclosures of gas insulated lines have a very low per unit length longitudinal resistance (see r_{en} in Table 3.5) due to a highest cross-sectional area: subsequently they are always solid-bonded so that an induced current almost equal to that of phase (but with opposite angle) can flow in it giving a strongest screening effect with very low additional power losses.

In this way there is an almost perfect decoupling between the phases so that a very good approximation consists in the study of GIL as a symmetric three-phase

Table 3.4 Typical data of 380 kV (50 Hz) GIL #5341 Al

Gas insulated line		
Cross-sectional area of phase (pipe) (Al IACS = 61%)	mm^2	5341
Cross-sectional area of enclosure (pipe) (Al alloy IACS = 52.57%)	mm^2	16 022
Phase outer diameter	mm	180
Phase thickness	mm	10
Enclosure inner diameter	mm	500
Enclosure thickness	mm	10
Insulating gas SF_6/N_2	%	20/80
Pressure	bar	7

* IACS = International Annealed Copper Standard (IACS = 100% ⇒ Copper Standard Conductivity = 58.108 m/(Ω mm^2))

Table 3.5 Positive sequence (50 Hz) parameters of 380 kV GIL #5341 Al

Installation	Gas insulated line Directly buried flat with axial spacing 1.3 m		
DC resistance of phase at 60 °C	r_{ph}	mΩ/km	6.286
DC resistance of enclosure at 50 °C	r_{en}	mΩ/km	2.330
Apparent resistance at positive sequence (50 Hz)	$r = r_{ph} + r_{en}$	mΩ/km	8.616
Phase-enclosure inductance	ℓ	mH/km	0.204
Phase-enclosure conductance	g	nS/km	negligible
Phase-enclosure capacitance	c	μF/km	0.0545
Longitudinal impedance (50 Hz)	$z = r_{ph} + r_{en} + j\omega\ell$	mΩ/km	8.6 + j 64
50 Hz shunt admittance	$y = g + j\omega c$	mS/km	0 + j 0.017
Characteristic impedance at 50 Hz	$\underline{Z}_c = \sqrt{\frac{\underline{z}}{\underline{y}}}$	Ω	61.46 ∠−0.07 rad
Propagation constant	$\underline{k} = \sqrt{\underline{z} \cdot \underline{y}}$ $\underline{k} = k' + j k''$	1/km	0.0001 + j 0.001
Natural power at 400 kV	$\frac{400^2}{\underline{Z}_c^*}$	MVA ∠rad	2603 ∠−0.07 rad
Capacitive current related to $U_0 = 400\,\text{kV}/\sqrt{3}$	I_{cap}	A/km	3.95
Steady state ampacity	I_a	A	2400

line with decoupled phase conductors having a very low apparent resistance equal to $r = r_{ph} + r_{en}$ (about the DC values).

Moreover, from Table 3.5 it is of note the low per unit length capacitance c = 54.5 nF/km and the high steady state ampacity (2400 A).

The capability charts of Figure 3.34 highlight immediately how a 100 km length GIL can have satisfying power flow performance with low power losses, good voltage levels at R, a satisfying behaviour for line energization (see U''_{0R}), the low I_{NL} no-load current in the line de-energization, even without shunt compensation.

GIL#5341 Al *d=100 km* I_c*=2400 A*	P_{RI}=1633 MW ΔP_l=14.7 MW ΔQ_l=*-161* Mvar	U''_{oR}=239.5 kV I_{NL}= 395 A Q_{NL}=-272.7 Mvar
	X''=20 Ω	ξ_{sh}=0.0

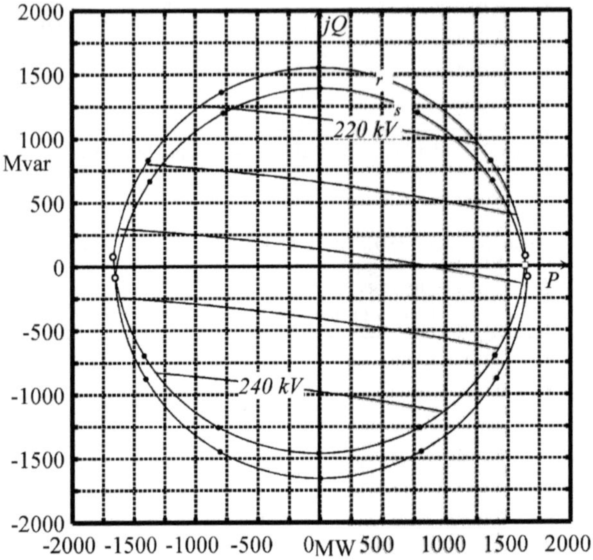

Figure 3.34 Capability chart for GIL; $d = 100\,\text{km}$

Moreover with the procedure in Section 3.9.2 it is possible to verify that, along the entire GIL length, the different regimes appear wholly satisfying.

3.11 Regimes with $U_{0S} \neq 230\,\text{kV}$

In the previous cases it has been always set $|\underline{U}_{0S}| = U_c = 230\,\text{kV}$ according to (3.3). Other capability charts with U_{0S} < or > 230 kV could be helpful to transmission system operators in order to optimize the network regimes in daily load variations.

3.12 "Receiving Area" and "Sending Area" as Set Intersection

The curvilinear lattices which embody the enhanced capability charts can be drawn by means of another type of procedure, which presents unquestionably the advantage of a noteworthy formal simplicity but has the drawback of not requiring a careful critical study on the possible interactions between the involved quantities of the network (it is usually very helpful to deeply understand the knowledge of the phenomena).

The abovementioned analyses which individuate and discriminate the steady state regimes compatible with the basic constraints can be also developed by means of the following attractive procedure (it is less circumstantial but equally valid) which is based on the set theory.

3.12.1 The Determination of the Receiving Area as Set Intersection

By utilizing (3.6)–(3.8) written in Section 3.3 for the first analysis and by not excluding the cases where the phasor $\underline{U}_{0S}$ determines the regimes with magnitudes I_S not allowable, it is immediate that the relation

$$\underline{S}_{R1} = 3 \cdot \left(\frac{\underline{U}_{0S}}{\underline{A}} - \frac{\underline{B} \cdot \underline{I}_R}{\underline{A}} \right) \cdot \underline{I}_R^* \tag{3.43}$$

(with $\delta = 0$–2π and $\underline{I}_R \equiv I_c \angle 0$) represents the enclosed contour which delimits the set of the receiving-end complex power compatible with the two constraints (3.1) and (3.3).

Then by using (3.12)–(3.15), written in Section 3.4 for the second analysis (and by not excluding the cases where the magnitudes of I_R become not allowable), the following relation

$$\underline{S}_{R2} = 3(\underline{A} \cdot \underline{U}_{0S} - \underline{B} \cdot \underline{I}_S) \cdot (-\underline{C} \cdot \underline{U}_{0S} + \underline{A} \cdot \underline{I}_S)^* \tag{3.44}$$

(with $\vartheta = 0$–2π and $\underline{I}_S \equiv I_c \angle 0$) represents the enclosed contour which delimits the set of the receiving-end complex powers compatible with the two constraints (3.2) and (3.3).

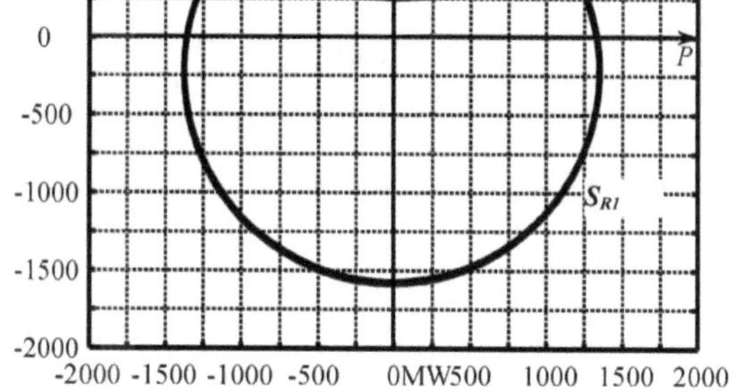

Figure 3.35 Receiving power area as intersection of sets (cable #b; 120 km; $\xi_{sh} = 0$) ($\underline{I}_R = \underline{I}_c = 1788$ A)

It can be easily understood how the contour delimiting the "receiving-end power area" of the regimes compatible simultaneously with the three "basic constraints" (3.1)–(3.3) can be obtained by means of an intersection procedure of the two sets above delimited as the example of Figure 3.35 clearly shows (in whole accordance with the capability charts of Figure 3.24).

It is important to highlight that the procedure allows identifying in automatic way also the contours of subsets corresponding to I_R and I_S inferior to I_c (e.g. for the drawing of the enhanced chart).

The structure of the given relations shows that the contours $\underline{S}_{R1}$ are circumferences and the contours $\underline{S}_{R2}$ are ellipses.

3.12.2 The Determination of the Sending Area as Set Intersection

By following some procedures completely analogous to those of the preceding paragraph, it is possible to achieve the following relation:

$$\underline{S}_{S1} = 3\underline{U}_{0S} \cdot \left(\frac{-\underline{C} \cdot \underline{U}_{0S}}{\underline{A}} + \frac{\underline{I}_R}{\underline{A}} \right)^* \tag{3.45}$$

with $\delta = 0\text{–}2\pi$ and $\underline{I}_R \equiv I_c \angle 0$ and the trivial one:

$$\underline{S}_{S2} = 3\underline{U}_{0S}\underline{I}_S^* \tag{3.46}$$

with $\vartheta = 0\text{–}2\pi$ and $\underline{I}_S \equiv I_c \angle 0$.

The intersection of the two sets gives the "sending power area".

3.13 The Analysis Along the Cable with Lumped Shunt Compensation

Once the investigations by means of the capability charts have been made (based on the hypothesis of distributed compensation), it is advisable to perform a further analysis by realistically considering a lumped shunt compensation (according with the criteria in Section 2.8.2) based on the same compensation degree ξ_{sh}, already evaluated with the idealized hypothesis; also in this new configuration it is possible to check that

i) the energization subtransient voltage $\underline{U}''_{0R}$ and the steady state current $\underline{I}_{NL}$ are practically within the limits already fixed;
ii) the receiving powers compatible with the ampacity constraints and the voltages along the whole link from the port S to the port R have slight variations.

Let us consider the example in Figure 3.21 (cable #b, $d = 60$ km, $\xi_{sh} = 0.608$ here repeated in Figure 3.36) where a lumped compensation is performed with shunt reactors installed every 20 km after the general rule recommended in Section 2.8.2, with the same compensation degree $\xi_{sh} = 0.608$.

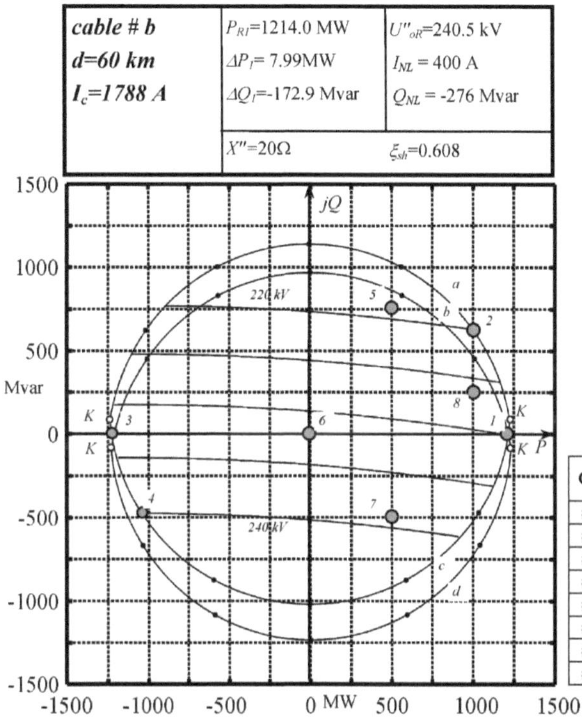

●	$\underline{S}_R$ MW+j Mvar
#1	1214 + j 0
#2	1000 + j 625
#3	-1214 - j 0
#4	-1044 - j 474
#5	500 + j 750
#6	0 + j 0
#7	500 - j 500
#8	1000 + j 250

Figure 3.36 Capability charts (cable #b; 60 km; $\xi_{sh} = 0.608$)

For the checks ii) it will be necessary (in correspondence to different meaningful complex powers, located along the border of the receiving area or inside it, as those indicated by the points 1–8 in the capability charts of Figure 3.36) to compute the current and voltage phasors, (or the hybrid vector $\underline{\boldsymbol{w}}_{x=0\text{–}19}$) in different ports along the lumped-compensated cable from R to S, and to visualize the diagrams of the magnitudes in Figures 3.37 and 3.38.

It can be realized by means of the matricial expression:

$$\underline{\boldsymbol{w}}_x = \underline{\boldsymbol{M}}_{xR} \cdot \underline{\boldsymbol{w}}_{\boldsymbol{R}} \,, \tag{3.47}$$

where each matrix $\underline{\boldsymbol{M}}_{xR}$ (transmission matrix between port x and port R), is given by the ordered product of the matrices of the system elements between x and R (see Figure 3.39).

For example (see Figure 3.39) it is:

$$\underline{\boldsymbol{M}}_{10R} = \underline{\boldsymbol{M}}_p \cdot \underline{\boldsymbol{M}}_p \cdot \underline{\boldsymbol{M}}_p \cdot \underline{\boldsymbol{M}}_{YY} \cdot \underline{\boldsymbol{M}}_p \cdot \underline{\boldsymbol{M}}_p \cdot \underline{\boldsymbol{M}}_p \cdot \underline{\boldsymbol{M}}_p \cdot \underline{\boldsymbol{M}}_p \cdot \underline{\boldsymbol{M}}_Y \,, \tag{3.48}$$

where:

$\underline{\boldsymbol{M}}_p$ is the transmission matrix of the cable circuit (uncompensated) of length equal to one-fifth the distance between a reactor and the following one;

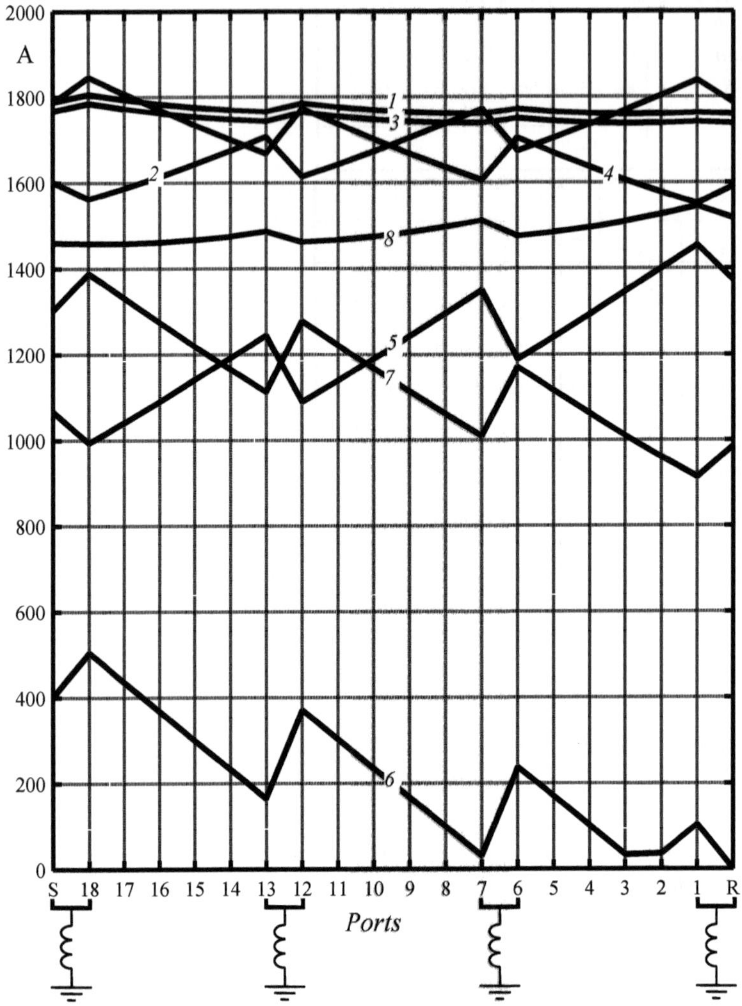

Figure 3.37 Steady state current magnitudes along the cable (cable #b; 60 km; $\xi_{\mathrm{sh}} = 0.608$, $U_{0S} = 230\,\mathrm{kV}$) with the lumped shunt compensation of Figure 3.39 in the regimes #1–8 of Figure 3.36

$\underline{\boldsymbol{M}}_Y$ and $\underline{\boldsymbol{M}}_{YY}$ are the shunt compensation matrices equal to those $\underline{\boldsymbol{M}}_\xi$ and $\underline{\boldsymbol{M}}_{\xi'}$ already presented in Section 2.8.2.

It is worth remembering that the hybrid vector $\underline{\boldsymbol{w}}_R$ is computable directly by means of Ossanna's method (already presented and used in Section 3.9.2) once fixed each receiving power from #1 to #8, the voltage level $\underline{U}_{0S}$ and the matrix $\underline{\boldsymbol{M}}_{SR}$.

The checks i) of page 82, regarding the subtransient regimes of energization at no-load and the steady state current $\underline{I}_{\mathrm{NL}}$, can be performed by extracting the parameters $\underline{A}$ and $\underline{B}$ from the matrix $\underline{\boldsymbol{M}}_{SR}$ and by applying successively (3.21)–(3.24): in

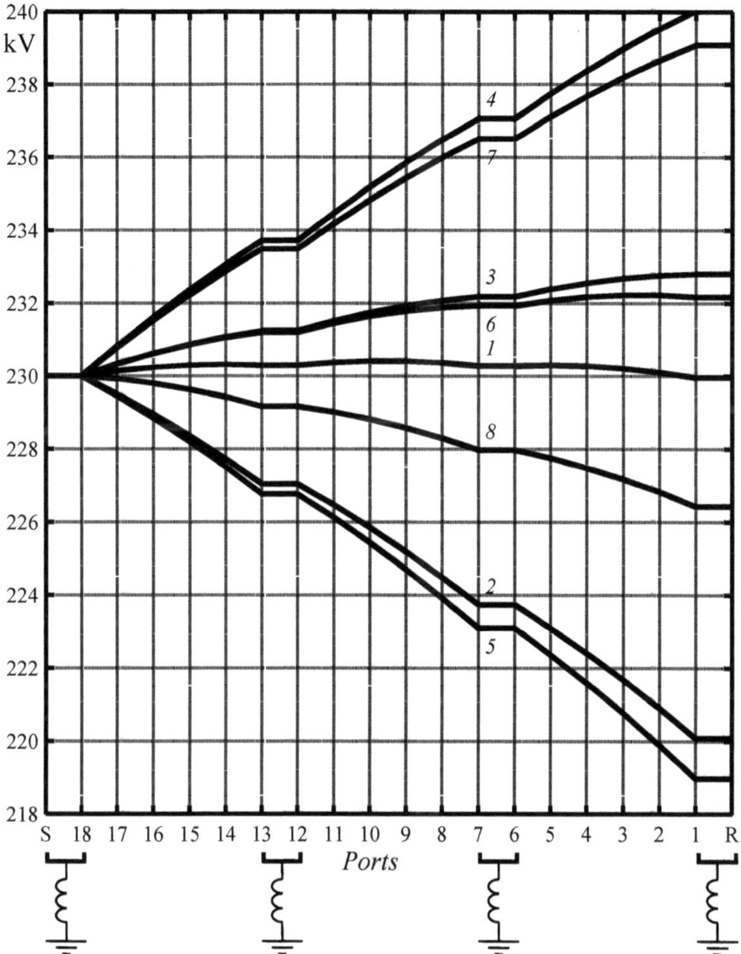

Figure 3.38 Steady state voltage magnitudes along the cable (cable #b; 60 km; $\xi_{sh} = 0.608$, $U_{0S} = 230\,\mathrm{kV}$) with the lumped shunt compensation of Figure 3.39 in the regimes #1–8 of Figure 3.36

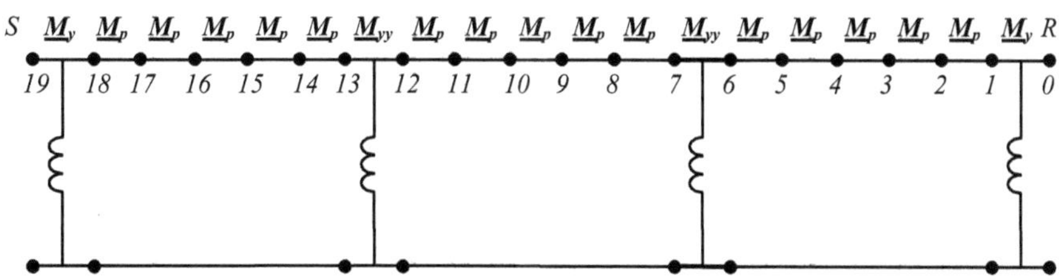

Figure 3.39 Composition of matrices for the analysis of lumped shunt compensation (cable #b; 60 km; $\xi_{sh} = 0.608$)

this way it yields (with $\xi_{sh} = 0.608$) the values $U''_{0R} = 240.55$ kV (with $X'' = 20\,\Omega$) and $I_{NL} = 401$ A (almost equal to the results given by distributed shunt compensation).

By means of other examples regarding steady state regimes #1–8, it can be ascertained that generally the hypothesis of ideal distributed shunt compensation is compatible with the real installation of lumped compensation: some differences in the receiving capability (e.g. with $\underline{S}_R$ #2 or $\underline{S}_R$ #4) gives some slight operation limitations.

The described procedure is obviously applicable to any configuration also very irregular for distances between reactors with different rated power.

3.14 Conclusions

This chapter gives efficient means in order to evaluate the power capability of EHV AC cables. The power capability charts represent an immediate and precise outlook on the possibilities of active and reactive power flows along the link. In the same charts, the regimes that give rise to receiving-end voltages too high or too low are highlighted. The method is based on the classic transmission formulae and has been fully developed in matrix form.

The procedure for the computation of the voltage and current magnitudes in any point along the cable has also been described. In many cases, the cable limit lengths can be overcome by means of shunt compensation: the capability charts show the different benefits depending upon the compensation degree.

The procedure has been also successfully applied to gas insulated transmission lines showing promising performance in power transmission.

References

1. R. Benato, A. Paolucci: Operating Capability of Long AC EHV Transmission Cables, *Electric Power Systems Research*, Vol. 75/1, July 2005, pp. 17–27
2. R. Arrighi in the name of WG 21.13: Operating Characteristics of long links of ac high voltage insulated cables, *Cigré 1986, Paper 21-13*, pp. 1–10
3. IEC 60287: Electric cables – Calculation of the current rating, (in 8 parts: 1.1, 1.2, 1.3, 2.1, 2.2, 3.1, 3.2, 3.3)
4. G.J. Anders: Rating of electric power cables: Ampacity computations for transmission, distribution, and Industrial Applications, IEEE Press, New York, 1997
5. G.J. Anders: Rating of electric power cables in unfavourable thermal environment, IEEE Press, Wiley-Interscience, New Jersey, 2005
6. S.Y. King, N.A. Halfter: Underground power cables, Longman, London and New York, 1982
7. E. Peschke, E. von Olshausen: Cable systems for high and extra-high voltage, 1999, Publicis MDC Verlag, Germany
8. IEC 62271-100 High voltage alternating-current circuit-breakers, 2001

9. British Electricity International: Modern Power Station Practice, EHV Transmission, Vol. K, Chapter 9, Sect. 3.3: Control of switching surges, Pergamon Press (Third Edition, 1991, Oxford)
10. W. Diesendorf: Insulation Coordination in High Voltage Electric Power Systems, Butterworths, London, 1974
11. C. Concordia: Synchronous Machines, J. Wiley & Sons, New York, 1951
12. E.W. Kimbark: Power system stability, J. Wiley & Sons, New York, 1948
13. J. Ossanna: Neue Arbeitsdiagramme über die Spannungsänderung in Wechselstromnetzen, *EuM*, 1926, pp. 113–126
14. J. Ossanna: Fernübertragungsmöglichkeiten großer Energiemengen, *ETZ*, 1922, pp. 1025–1029, 1061–1063

Chapter 4
Operating Capability of AC EHV Mixed Lines with Overhead and Cables Links

4.1 Introduction

The previous chapter has been devoted to the operating capability and constraints involved in power transmission of long AC cable links.

In this chapter the approach has been extended in order to show the transmission performance of some typical mixed configurations with overhead lines (OHL) and underground cable lines (UGC).

In order to erect a new long link, (but also, often, to fit an existing one due to new territory necessities) it could be necessary to have a mix of transmission technologies such as a cascade connection of OHL–UGC–OHL. In fact, the overhead line can be unsuitable in certain parts of the route for tackling some environmental and orographical problems, which can be easily overcome with (as already stated in Chapter 3) UGCs.

For example, an UGC section allows a HV/EHV link to pass inside galleries or through areas too wide for OHL span such as large lakes or arms of the sea: it is sometimes called a "siphon". This configuration may also solve some critical issues, due to strong local oppositions or some laws, since it drastically reduces the environmental and magnetic impacts and hence it permits the UGC line to pass through or near a protected site (sensible places such as a school or natural park) or an urbanized area also with economical convenience (see Chapter 6). It is worth remembering that some important installations of mixed OHL–UGC links have been realized for several years to allow great power flows towards town centres. It must be emphasized that the use of UGCs gives usually the economic burden of the shunt compensation reactors.

For the aforementioned reasons, the transmission system operators will have to face more and more the analysis of mixed lines aiming at achieving their possible performances without jeopardizing the system safety and the power quality.

4.2 Mixed Lines: OHL–UGC–OHL

A rigorous approach to the issue requires to fix, for each OH and UG section, the specified constraints regarding the safety and the lifetime of section itself and to research successively the maximum compatible performances of the whole link. A simpler approach [1] can be adopted when the situation is that of Figure 4.1 (① OHL–② UGC–③ OHL, with lengths d_1, d_2, and d_3), where the intermediate section is constituted of UGC: in this (very usual) case the steady state constrains expressed by (4.1) and (4.2), analogous to (3.1) and (3.2), appear sufficient to satisfy the ampacity not only of the cable but also of OHL ① and ③:

$$|\underline{I}_H| \leq I_c\,, \tag{4.1}$$
$$|\underline{I}_K| \leq I_c\,, \tag{4.2}$$
$$|\underline{U}_{0S}| = U_{0c}\,. \tag{4.3}$$

the value of I_c represents a current level which must not exceed, with a conservative margin, the steady state ampacity of cable section ②: it is sufficient that I_c (imposed at H and K) is little lower than the ampacity of OHL ① and ③ to obtain (as it will have to be systematically verified) allowable current levels also at S and R. Ultimately, (4.3) fixes the voltage at S, similarly to the settlement already adopted in Chapter 3. Also in the following treatment, where $U_m = 420\,\text{kV}$, the level $U_{0c} = 230\,\text{kV} = 95\%\, U_m/\sqrt{3}$ has been considered.

For each line section a suitable "positive sequence model" has been assumed: this requires phase transpositions for OHL and for cable line also the cross-bonding arrangement and allows the use of classical transmission equations based on positive sequence parameters (see Table 4.1 for OHL and Table 4.2 for single-circuit UGC) and of relevant transmission matrices (see Chapter 2).

This is a useful guide also for little unbalanced systems provided that a slightly conservative choice of I_c and U_{0c} is made. Successively, a deep insight can be performed by means of multiconductor study similarly to Chapter 5 [2–4].

In the case hereafter considered, for OHL a typical 380 kV line of Italian grid (equipped with bundles of three sub-conductors) has been assumed and for UGC ② almost always a 380 kV double-circuit consisting of single-pole 2500 mm^2 copper conductor XLPE cables (in the same configuration as in Figure 6.2 of Chapter 6): by considering the wide circuit spacing (5 m), the two single-circuits can be assumed thermally and electrically independent so that, for the double-circuit, the ampacity is 2 × 1.788 A and the electrical model is computable after Section 2.7.

Once considered the values of ampacity for OHL (foreseen by Italian Standard; see Table 4.1), the levels I_c (during cold months) = 2800 A and I_c (during warm

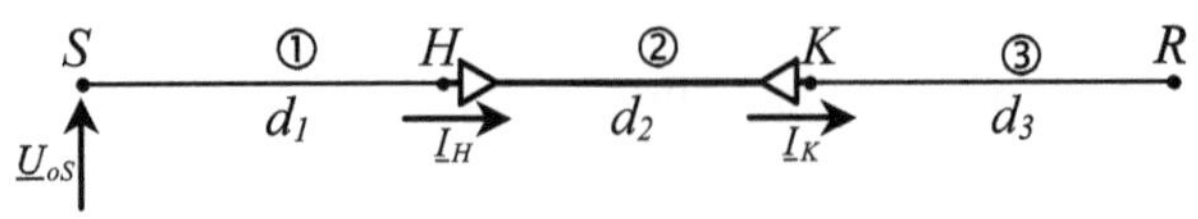

Figure 4.1 Typical sketch of a mixed line

months) = 2100 A have been assumed in (4.1) and (4.2), as specified in each case studies. These choices do not seem to be excessively redundant for section ② with $I_a = 2 \cdot 1788$ A as shown in Table 4.2: ($\frac{2\cdot 1788}{2800} = 1.28$ during cold months and $\frac{2\cdot 1788}{2100} = 1.70$ during warm months) because it seems a conservative installation criterion to have a good power reserve in case of a circuit failure.

The analysis of the link between nodes S and R, not inserted in the real context of a given electric network, constitutes a preliminary assessment and a useful guide for further network studies (among which the power flows and network simulations are indispensable).

In analogy with Chapter 3, the regime of this mixed line can be achieved by means of a first analysis ($\underline{U}_{0S}$ and $|\underline{I}_K|$ constrained) and successively a second one ($\underline{U}_{0S}$ and $|\underline{I}_H|$ constrained), with the aim at individuating the regimes that respect the constraints (4.1)–(4.3).

With regard to the steady state capacitive power Q_C absorbed by the UGC considered in this chapter (a double circuit of 2500 mm^2 cables) it can be roughly estimated (refer to Section 3.2) $Q_C \approx 22$ Mvar/km at 50 Hz and ≈ 26 Mvar/km at 60 Hz: it requires almost always a suitable degree of shunt compensation ξ_{sh} both to improve the steady state and transient regimes (see Section 4.7).

Table 4.1 Positive sequence parameters (50 Hz) of OHL with bundles of 3 ACSR sub-conductors, (Φ 31.5 mm, spacing 400 mm)

Overhead section ① and ③			
Voltage level			380 kV
Resistance at 75 °C (50 Hz)	r	mΩ/km	23.10
Series inductance	ℓ	mH/km	0.858
Shunt leakance (50 Hz)	g	nS/km	10
Shunt capacitance	c	μF/km	0.0133
Ampacity (cold months) – Italian standard	I_a	A	2955
Ampacity (warm months) – Italian standard			2220

Table 4.2 (Cp. Table 3.2) Positive sequence parameters (50 Hz) of single-circuit UGC with single-core cables directly buried in cross bonding at a depth 1.45 m in the soil with $\rho_{th} = 1.0\,\mathrm{K\cdot m/W}$; spacing $s = 0.35$ m; flat laying and soil ambient temperature $\vartheta_a = 20\,°\mathrm{C}$

Cable section ②			
Voltage level			380 kV
Phase cross-section		mm^2	2500 Cu
Phase resistance at 90 °C	r	mΩ/km	13.3
Series inductance	ℓ	mH/km	0.576
Shunt leakance with $\tan\delta = 0.0007$	g	nS/km	52
Shunt capacitance with $\varepsilon_r = 2.3$	c	μF/km	0.234
Steady state ampacity	I_a	A	1788

Nevertheless, the value of ξ_{sh} does not modify the methodology for the first and second analysis in order to construct the capability charts (presented in Sections 4.4–4.6) so that it is more convenient to calculate it definitively in Section 4.7.

In any case, the hypothesis of uniformly distributed shunt compensation can be adopted also in this chapter (analogously to Chapter 3): it allows constructing easily the corresponding transmission matrix to introduce in the system analysis.

4.3 The Transmission Matrices for the System Study

For each section ①, ② and ③ and for their cascade composition, the corresponding classical matrices can be derived as in Section 2.6.

It is worth underlining that in the following case studies the matrices $\underline{\boldsymbol{M}}_1$ and $\underline{\boldsymbol{M}}_3$ (OHL) must be computed by considering the corresponding lengths d_1 and d_3; the matrix $\underline{\boldsymbol{M}}_2$ must be computed by considering both the length d_2 and the compensation degree ξ_{sh} determined as in Section 4.7.

4.4 First Analysis

Once fixed at node K the current phasor $\underline{I}_K = I_c\angle 0$ (on the real axis), and imposed at node S the following phasor

$$\underline{U}_{0S} = 230\,\mathrm{kV}\angle\delta \quad (\delta = 0\text{–}2\pi)\ ,$$

a set of regimes of the entire line is univocally achieved, among which only those respecting the limit I_c at node H must be individuated. In particular (4.4)–(4.11), show the subsequent formal passages, which give (4.13), deriving from (4.11). The expression (4.13) can be interpreted by the phasorial diagram of Figure 4.2 (with some modified proportions for graphical purposes) which, by means of the relations shown in the same figure, yields the problem solution: for δ ranging between δ_1 and δ_2 (evaluated by means of (4.14) and (4.15)), the voltage phasor $\underline{U}_{0S}(\delta)$ determines in this first analysis all the current phasors $\underline{I}_H$ (e.g. OL_1, OL_3, OL_4, OL_2) having magnitudes within the limit I_c according to (4.2) as long as the magnitudes of phasors $\underline{a}$, $\underline{b}$, $\underline{c}$ are compatible with the triangular configuration. In particular, the phasors OL_1 and OL_2, (the former leading and the latter lagging with respect to $\underline{I}_K$) regard currents $\underline{I}_H$ with the same magnitude of $I_c = |\underline{I}_K|$.

$$\underbrace{\begin{bmatrix} \underline{U}_{0S} \\ \underline{I}_S \end{bmatrix}}_{\underline{v}_{SS(K)}} = \underbrace{\begin{bmatrix} \underline{A}_{12} & \underline{B}_{12} \\ \underline{C}_{12} & \underline{D}_{12} \end{bmatrix}}_{\underline{\boldsymbol{M}}_{12}} \underbrace{\begin{bmatrix} \underline{U}_{0K} \\ \underline{I}_K \end{bmatrix}}_{\underline{v}_K}\ , \tag{4.4}$$

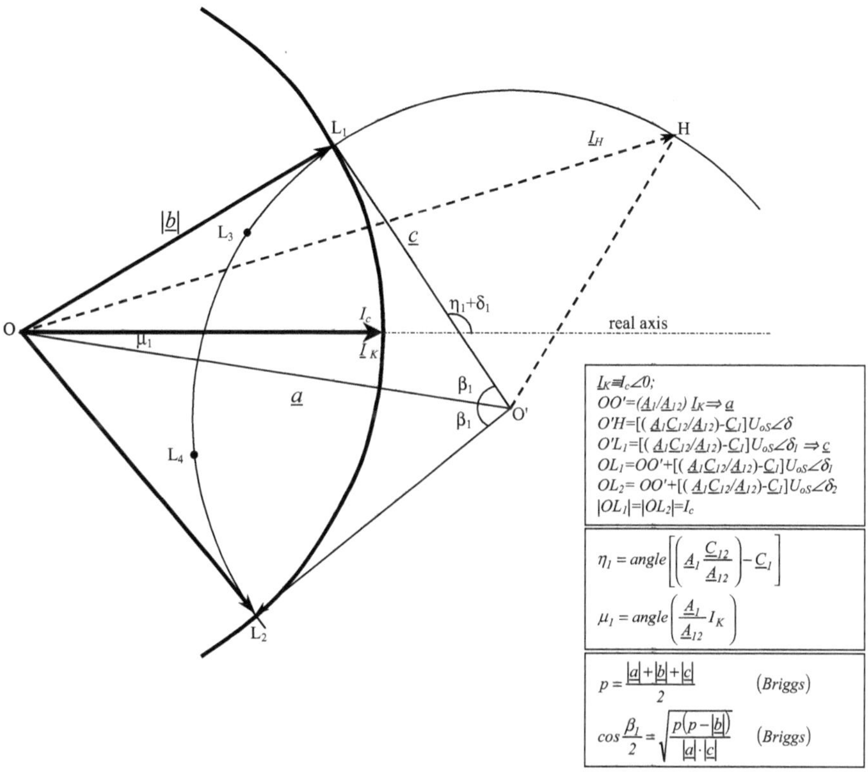

Figure 4.2 Phasorial diagram of (4.13)

$$\underbrace{\begin{bmatrix} \underline{U}_{0K} \\ \underline{I}_K \end{bmatrix}}_{\underline{\boldsymbol{v}}_K} = \underbrace{\begin{bmatrix} \underline{D}_{12} & -\underline{B}_{12} \\ -\underline{C}_{12} & \underline{A}_{12} \end{bmatrix}}_{\underline{\boldsymbol{M}}_{12}^{-1}} \underbrace{\begin{bmatrix} \underline{U}_{0S} \\ \underline{I}_S \end{bmatrix}}_{\underline{\boldsymbol{v}}_{SS(K)}}, \tag{4.5}$$

$$\underbrace{\begin{bmatrix} \underline{U}_{0S} \\ \underline{I}_K \end{bmatrix}}_{\underline{\boldsymbol{v}}_{SK}} = \begin{bmatrix} 1 & 0 \\ -\underline{C}_{12} & \underline{A}_{12} \end{bmatrix} \underbrace{\begin{bmatrix} \underline{U}_{0S} \\ \underline{I}_S \end{bmatrix}}_{\underline{\boldsymbol{v}}_{SS(K)}}, \tag{4.6}$$

$$\underline{\boldsymbol{N}}_{S1} = \begin{bmatrix} 1 & 0 \\ -\underline{C}_{12} & \underline{A}_{12} \end{bmatrix}^{-1}, \tag{4.7}$$

$$\underbrace{\begin{bmatrix}\underline{U}_{0S}\\ \underline{I}_S\end{bmatrix}}_{\underline{v}_{SS(K)}} = \underbrace{\begin{bmatrix}1 & 0\\ \frac{\underline{C}_{12}}{\underline{A}_{12}} & \frac{1}{\underline{A}_{12}}\end{bmatrix}}_{\underline{N}_{S1}} \underbrace{\begin{bmatrix}\underline{U}_{0S}\\ \underline{I}_K\end{bmatrix}}_{\underline{v}_{SK}}, \qquad \blacksquare: \ \text{imposed phasors}, \tag{4.8}$$

$$\underline{v}_{HH(K)} = \underline{M}_1^{-1} \cdot \underline{v}_{SS(K)} \quad \Rightarrow \quad \underline{v}_{HH(K)} = \underline{M}_1^{-1} \cdot \underline{N}_{S1} \cdot \underline{v}_{SK}, \tag{4.9}$$

$$\underbrace{\begin{bmatrix}\underline{U}_{0H}\\ \underline{I}_H\end{bmatrix}}_{\underline{v}_{HH(K)}} = \underbrace{\begin{bmatrix}\underline{D}_1 & -\underline{B}_1\\ -\underline{C}_1 & \underline{A}_1\end{bmatrix}}_{\underline{M}_1^{-1}} \underbrace{\begin{bmatrix}1 & 0\\ \frac{\underline{C}_{12}}{\underline{A}_{12}} & \frac{1}{\underline{A}_{12}}\end{bmatrix}}_{\underline{N}_{S1}} \underbrace{\begin{bmatrix}\underline{U}_{0S}\\ \underline{I}_K\end{bmatrix}}_{\underline{v}_{SK}}, \tag{4.10}$$

$$\underbrace{\begin{bmatrix}\underline{U}_{0H}\\ \underline{I}_H\end{bmatrix}}_{\underline{v}_{HH(K)}} = \underbrace{\begin{bmatrix}\underline{D}_1 - \underline{B}_1\frac{\underline{C}_{12}}{\underline{A}_{12}} & -\frac{\underline{B}_1}{\underline{A}_{12}}\\ -\underline{C}_1 + \underline{A}_1\frac{\underline{C}_{12}}{\underline{A}_{12}} & \frac{\underline{A}_1}{\underline{A}_{12}}\end{bmatrix}}_{\underline{N}_{H1}} \underbrace{\begin{bmatrix}\underline{U}_{0S}\\ \underline{I}_K\end{bmatrix}}_{\underline{v}_{SK}}, \tag{4.11}$$

$$\underline{N}_{H1} = \underline{M}_1^{-1}\underline{N}_{S1}, \tag{4.12}$$

$$\underline{I}_H = \left(-\underline{C}_1 + \frac{\underline{A}_1 \cdot \underline{C}_{12}}{\underline{A}_{12}}\right) \cdot \underline{U}_{0S} + \frac{\underline{A}_1}{\underline{A}_{12}} \cdot \underline{I}_K, \tag{4.13}$$

$$\delta_1 = \pi - \beta_1 - \eta_1 + \mu_1 \quad (\text{where } \mu_1 < 0), \tag{4.14}$$

$$\delta_2 = \delta_1 + 2\beta_1 . \tag{4.15}$$

4.5 Second Analysis

In an analogue way, once fixed at node H the current phasor $\underline{I}_H = I_c\angle 0$ (on the real axis), and imposed at node S the following phasors

$$\underline{U}_{0S} = 230\,\text{kV}\angle\vartheta \quad (\vartheta = 0\text{–}2\pi),$$

another set of regimes can be achieved among which only those respecting the limit I_c at node K are of interest.

The relations (4.16)–(4.23) show the subsequent formal passages, which give rise to (4.25). It can be interpreted by the phasorial diagram of Figure 4.3: by means of the relations shown in the same figure, it is possible to see that, for ϑ ranging

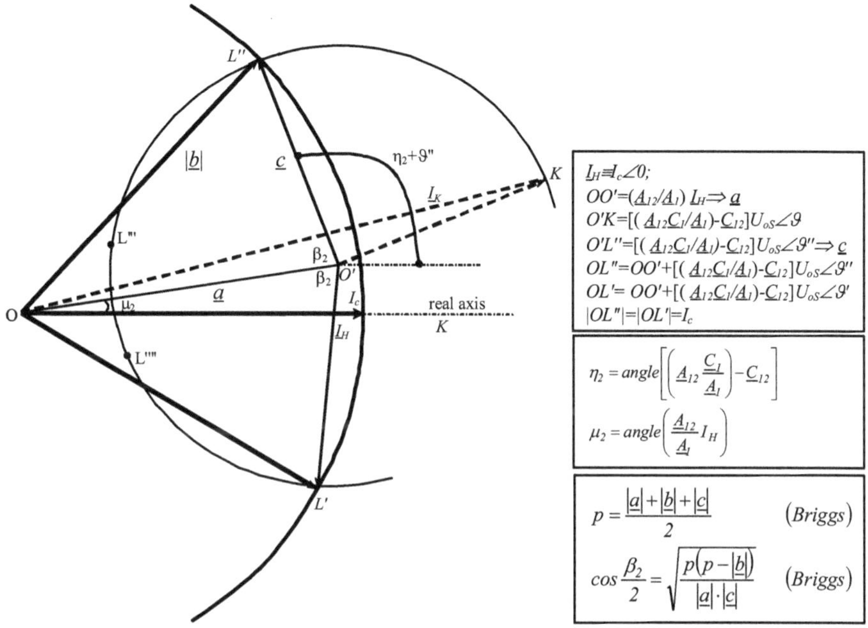

Figure 4.3 Phasorial diagram of (4.25)

between ϑ'' and ϑ' (evaluated by means of (4.26) and (4.27)), the voltage phasor $\underline{U}_{0S}(\vartheta)$ determines in this second analysis all the current phasors $\underline{I}_K$ (e.g. OL'', OL''', OL'''', OL') with magnitudes within the limit I_c, according to (4.1).

$$\underbrace{\begin{bmatrix} \underline{U}_{0S} \\ \underline{I}_S \end{bmatrix}}_{\underline{\boldsymbol{v}}_{SS(H)}} = \underbrace{\begin{bmatrix} \underline{A}_1 & \underline{B}_1 \\ \underline{C}_1 & \underline{D}_1 \end{bmatrix}}_{\underline{\boldsymbol{M}}_1} \underbrace{\begin{bmatrix} \underline{U}_{0H} \\ \underline{I}_H \end{bmatrix}}_{\underline{\boldsymbol{v}}_H}, \qquad (4.16)$$

$$\underbrace{\begin{bmatrix} \underline{U}_{0H} \\ \underline{I}_H \end{bmatrix}}_{\underline{\boldsymbol{v}}_H} = \underbrace{\begin{bmatrix} \underline{D}_1 & -\underline{B}_1 \\ -\underline{C}_1 & \underline{A}_1 \end{bmatrix}}_{\underline{\boldsymbol{M}}_1^{-1}} \underbrace{\begin{bmatrix} \underline{U}_{0S} \\ \underline{I}_S \end{bmatrix}}_{\underline{\boldsymbol{v}}_{SS(H)}}, \qquad (4.17)$$

$$\underbrace{\begin{bmatrix} \underline{U}_{0S} \\ \underline{I}_H \end{bmatrix}}_{\underline{\boldsymbol{v}}_{SH}} = \begin{bmatrix} 1 & 0 \\ -\underline{C}_1 & \underline{A}_1 \end{bmatrix} \underbrace{\begin{bmatrix} \underline{U}_{0S} \\ \underline{I}_S \end{bmatrix}}_{\underline{\boldsymbol{v}}_{SS(H)}}, \qquad (4.18)$$

$$\underline{N}_{S2} = \begin{bmatrix} 1 & 0 \\ -\underline{C}_1 & \underline{A}_1 \end{bmatrix}^{-1}, \tag{4.19}$$

$$\underbrace{\begin{bmatrix} \underline{U}_{0S} \\ \underline{I}_S \end{bmatrix}}_{\underline{v}_{SS(H)}} = \underbrace{\begin{bmatrix} 1 & 0 \\ \frac{\underline{C}_1}{\underline{A}_1} & \frac{1}{\underline{A}_1} \end{bmatrix}}_{\underline{N}_{S2}} \underbrace{\begin{bmatrix} \underline{U}_{0S} \\ \underline{I}_H \end{bmatrix}}_{\underline{v}_{SH}}, \tag{4.20}$$

$$\underline{v}_{KK(H)} = \underline{M}_{12}^{-1} \cdot \underline{v}_{SS(H)} \quad \Rightarrow \quad \underline{v}_{KK(H)} = \underline{M}_{12}^{-1} \cdot \underline{N}_{S2} \cdot \underline{v}_{SH}, \tag{4.21}$$

$$\underbrace{\begin{bmatrix} \underline{U}_{0K} \\ \underline{I}_K \end{bmatrix}}_{\underline{v}_{KK(H)}} = \underbrace{\begin{bmatrix} \underline{D}_{12} & -\underline{B}_{12} \\ -\underline{C}_{12} & \underline{A}_{12} \end{bmatrix}}_{\underline{M}_{12}^{-1}} \underbrace{\begin{bmatrix} 1 & 0 \\ \frac{\underline{C}_1}{\underline{A}_1} & \frac{1}{\underline{A}_1} \end{bmatrix}}_{\underline{N}_{S2}} \underbrace{\begin{bmatrix} \underline{U}_{0S} \\ \underline{I}_H \end{bmatrix}}_{\underline{v}_{SH}}, \tag{4.22}$$

$$\underbrace{\begin{bmatrix} \underline{U}_{0K} \\ \underline{I}_K \end{bmatrix}}_{\underline{v}_{KK(H)}} = \underbrace{\begin{bmatrix} \underline{D}_{12} - \underline{B}_{12}\frac{\underline{C}_1}{\underline{A}_1} & -\frac{\underline{B}_{12}}{\underline{A}_1} \\ -\underline{C}_{12} + \underline{A}_{12}\frac{\underline{C}_1}{\underline{A}_1} & \frac{\underline{A}_{12}}{\underline{A}_1} \end{bmatrix}}_{\underline{N}_{K2}} \underbrace{\begin{bmatrix} \underline{U}_{0S} \\ \underline{I}_H \end{bmatrix}}_{\underline{v}_{SH}}, \tag{4.23}$$

$$\underline{N}_{K2} = \underline{M}_{12}^{-1} \underline{N}_{S2}, \tag{4.24}$$

$$\underline{I}_K = \left(-\underline{C}_{12} + \frac{\underline{A}_{12} \cdot \underline{C}_1}{\underline{A}_1} \right) \cdot \underline{U}_{0S} + \frac{\underline{A}_{12}}{\underline{A}_1} \cdot \underline{I}_H, \tag{4.25}$$

In particular, the phasors OL'' and OL' reproduce again the same relations of magnitude and angle (with respect to $|\underline{I}_H|$) detectable in the first analysis for OL_1 and OL_2.

$$\vartheta'' = \pi - \beta_2 - \eta_2 + \mu_2 \quad (\text{where } \mu_2 > 0), \tag{4.26}$$

$$\vartheta' = \vartheta'' + 2\beta_2 . \tag{4.27}$$

4.6 The Capability Charts

The procedures presented in the first and second analysis allow individuating, for a given mixed line, all the regimes at power frequency compatible with the constraints (4.1)–(4.3), yielding for each regime the vector (composed of voltage and current phasors) at each node: it becomes particularly interesting to evaluate the complex (three-phase) power $\underline{S}_S$ and $\underline{S}_R$ by using respectively the elements of vectors $\underline{\boldsymbol{v}}_{SS(K)}$ in (4.8) and $\underline{\boldsymbol{v}}_{SS(H)}$ in (4.20) and the elements of vectors $\underline{\boldsymbol{v}}_{RR(K)}$ in (4.28) and $\underline{\boldsymbol{v}}_{RR(H)}$ in (4.29).

For the explicit form of $\underline{\boldsymbol{N}}_{R1}$ and $\underline{\boldsymbol{N}}_{R2}$ refer to Section 4.9.

It is worth noting that (4.28) and (4.29) as well as (4.8) and (4.20) are useful not only to compute the end port power but also to verify (as already premised in Section 4.2) the respect of ampacity at R and S for OHLs.

$$\underbrace{\begin{bmatrix} \underline{U}_{0R} \\ \underline{I}_R \end{bmatrix}}_{\underline{\boldsymbol{v}}_{RR(K)}} = \underbrace{\underbrace{\begin{bmatrix} \underline{D}_{123} & -\underline{B}_{123} \\ -\underline{C}_{123} & \underline{A}_{123} \end{bmatrix}}_{\underline{\boldsymbol{M}}_{123}^{-1}} \underbrace{\begin{bmatrix} 1 & 0 \\ \frac{\underline{C}_{12}}{\underline{A}_{12}} & \frac{1}{\underline{A}_{12}} \end{bmatrix}}_{\underline{\boldsymbol{N}}_{S1}}}_{\underline{\boldsymbol{N}}_{R1}} \underbrace{\begin{bmatrix} \underline{U}_{0S} \\ \underline{I}_K \end{bmatrix}}_{\underline{\boldsymbol{v}}_{SK}}, \tag{4.28}$$

$$\underbrace{\begin{bmatrix} \underline{U}_{0R} \\ \underline{I}_R \end{bmatrix}}_{\underline{\boldsymbol{v}}_{RR(H)}} = \underbrace{\underbrace{\begin{bmatrix} \underline{D}_{123} & -\underline{B}_{123} \\ -\underline{C}_{123} & \underline{A}_{123} \end{bmatrix}}_{\underline{\boldsymbol{M}}_{123}^{-1}} \underbrace{\begin{bmatrix} 1 & 0 \\ \frac{\underline{C}_{1}}{\underline{A}_{1}} & \frac{1}{\underline{A}_{1}} \end{bmatrix}}_{\underline{\boldsymbol{N}}_{S2}}}_{\underline{\boldsymbol{N}}_{R2}} \underbrace{\begin{bmatrix} \underline{U}_{0S} \\ \underline{I}_H \end{bmatrix}}_{\underline{\boldsymbol{v}}_{SH}}. \tag{4.29}$$

The respect of ampacity along the UGC length appears (in any case-study) assured a priori, since the ratios between ampacity and I_c are 1.28 (cold months) and 1.7 (warm months). If the ratio were less conservative, it would be advisable to make some checks by means of Section 4.10.

Hence the outlines can be drawn (see the example of Figure 4.4): the outline s bounds the "sending-end power area" and the outline r the "receiving-end power area", respecting the fixed constraints. It can be ascertained (see Section 4.9) that they are composed of arcs of circumference and ellipse, analytically inferable.

At regular steps, it is possible to mark the complex power $\underline{S}_S$ and $\underline{S}_R$ with couple of points upon the outlines s and r: their correspondence is easily detectable by the mutual proximity (see Figure 4.4). The difference $\underline{S}_S - \underline{S}_R$ views with the real part the active transmission power losses and with the imaginary part the reactive power demand by the mixed transmission system (including reactive compensation of UGC, if any).

The couples of points marked with $\times$ represent the particular regimes characterized by $|\underline{I}_K| \equiv |\underline{I}_H| \equiv I_c$.

These capability charts for mixed link give (as already seen for only cable lines) immediate visual information on power flow performances.

As shown in the first example of Figure 4.4, each capability chart is completed with the following transmission parameters analogously to Chapter 3:

Figure 4.4 A first example of capability charts

d_1=60 km	P_{R1}=1850 MW	U''_{0R}=242.5 kV
d_2=30 km	ΔP_1=48.5 MW	I_{NL}= 391 A
d_3=20 km	ΔQ_1=323 Mvar	Q_{NL}=-270 Mvar
I_c=2800 A	X''=15 Ω	ξ_{sh}=0.7 (*see § 4.7*)

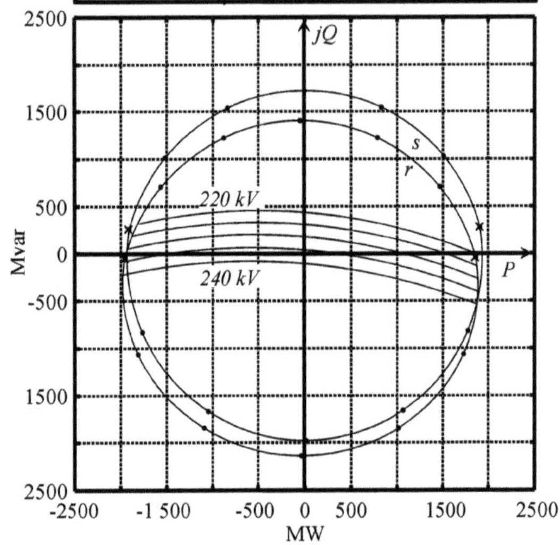

d_1, d_2, d_3	lengths of sections ①, ② and ③;
I_c	chosen current constraint at H and K;
X''	50 Hz subtransient reactance (seen at port S);
ξ_{sh}	shunt compensation degree (for section ② according to Section 4.7);
U''_{0R}	RMS value of the subtransient voltage in the no-load mixed line energization from the port S for the given degree ξ_{sh};
I_{NL}	RMS value of the current absorbed in S by the mixed line at no load in R, in steady state regime with $\lvert \underline{U}_{0S} \rvert = 230\,\text{kV}$ and with the given degree ξ_{sh};
Q_{NL}	Capacitive reactive power absorbed in S by the mixed link at no load in R in steady state regime with $\lvert \underline{U}_{0S} \rvert = 230\,\text{kV}$ and with the given ξ_{sh};
P_{R1}	maximum active receiving power with $\cos\varphi = 1$;
ΔP_1	active power losses in the mixed line at P_{R1};
ΔQ_1	whole reactive power required by the mixed line at P_{R1}.

The maximum active receiving-end power P_{R1} (at $\cos\varphi = 1$) is generally detectable in the around of δ_1 in the first analysis or in the around of ϑ' in the second analysis and appears very meaningful in order to evaluate the more interesting performances and power flows of the mixed line, together with the active losses ΔP_1 and the reactive power ΔQ_1 required by the mixed line.

4.6.1 Phase Voltage Levels at R

The voltage levels at node R play a key role in the network operation and can be directly visualized in the "receiving-end power area" analogously to Section 3.6: in this case the following formula must be applied

$$\underline{S}_R = 3\underline{U}_{0R}[(\underline{U}_{0S} - \underline{A}_{123}\underline{U}_{0R})/\underline{B}_{123}]^* \ . \tag{4.30}$$

4.7 No-Load Energization and De-Energization

In the mixed lines the energization and de-energization phenomena deserve careful consideration also in the cases where the cable length d_2 is small, since the inductive reactance of the section OHL, which adds to X'', tends to heighten, in no load regimes, the whole capacitive effect.

The sketch of Figure 4.5 shows the switch-on of the circuit breaker in S in order to energize the mixed line.

As already seen in Section 3.7, the supply in S can be modelled as an equivalent generator which is characterized by its electromotive force U_0 (supposed to be equal to 230 kV) and the short-circuit subtransient impedance $\underline{Z}''$ (for simplicity purely inductive $\underline{Z}'' = \mathrm{j}\,X''$).

The power frequency subtransient voltage $\underline{U}''_{0R}$, due to the closing at S and with R open, is completely defined by the following formulae:

$$\underline{U}''_{0S} = \left(\frac{U_0}{\underline{Z}'' + \frac{\underline{A}_{123}}{\underline{C}_{123}}}\right) \cdot \frac{\underline{A}_{123}}{\underline{C}_{123}} \ , \tag{4.31}$$

$$\underline{U}''_{0R} = \frac{\underline{U}''_{0S}}{\underline{A}''_{123}} \ , \tag{4.32}$$

$$\underline{U}''_{0R} = \frac{U_0}{\underline{A}_{123} + \underline{Z}'' \cdot \underline{C}_{123}} \ , \tag{4.33}$$

being $\underline{A}_{123}/\underline{C}_{123}$ the impedance (almost completely capacitive) as seen from S with R open, and expressing with (4.32) the Ferranti's effect; for the subtransient reactance X'' values ranging in the interval 4–23 Ω can be assumed as already seen in Section 3.7.

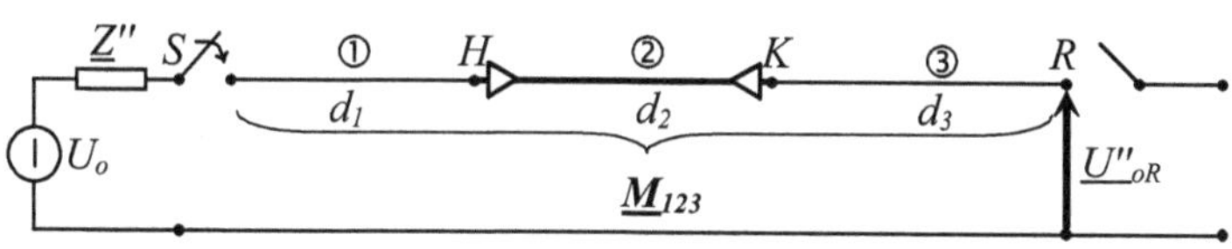

Figure 4.5 Typical sketch for energization of mixed line at no-load at R (subtransient regime)

In the hypothesis that, extinguished the transient switching phenomena, the network voltage regulation restores again at port S the rated value $U_0 = 230\,\text{kV}$, the no-load steady state (almost entirely capacitive) current which the circuit breaker must be able to interrupt is computed by

$$\underline{I}_{\text{NL}} = \frac{U_0}{(\underline{A}_{123}/\underline{C}_{123})}\,. \tag{4.34}$$

With regard to the energization phenomena, in order to respect the standard switching levels (e.g. 1050 kV) with a conservative margin, it is advisable (as in Section 3.7) to impose the following constraint to (4.33)

$$|\underline{U}''_{0R}| \leq 242.5\,\text{kV} = U_{\text{m}}/\sqrt{3}\,. \tag{4.35}$$

As regards the de-energization phenomena, the following constraint to (4.34) must be imposed:

$$|\underline{I}_{\text{NL}}| \leq 400\,\text{A} \tag{4.36}$$

which expresses the suggested limit of the already mentioned standard.

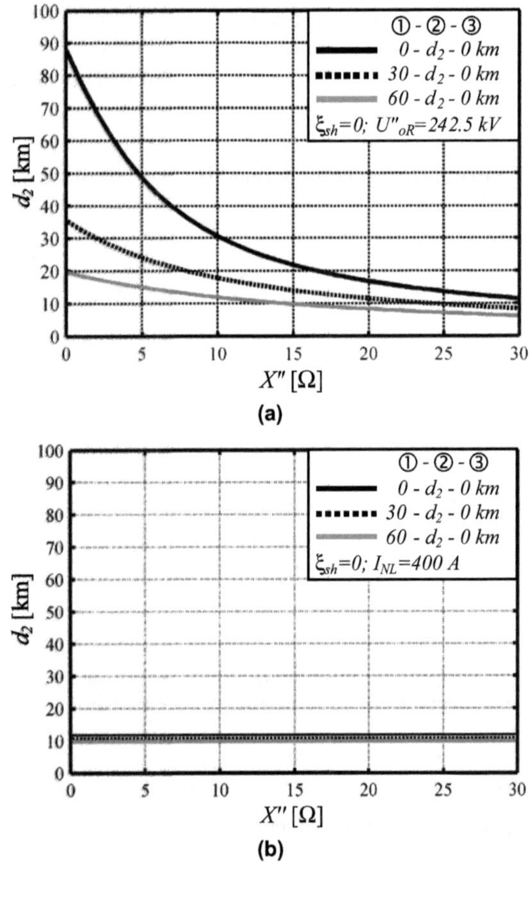

Figure 4.6 Examples of limit length d_2 for uncompensated UGC; (**a**) due to the constraint (4.35); (**b**) due to the constraint (4.36)

Therefore, once fixed the parameters d_1, d_2, d_3, X'', U_0, the values of ξ_{sh} which give the respect of both the abovementioned constraints can easily be computed with automatic iterations of (4.33) and (4.34).

In the numerous examples presented in the following Section 4.8, it will be possible to distinguish which constraint (4.35) or (4.36) is decisive for the choice of the shunt compensation degree, depending on the different configurations.

Since the determination of ξ_{sh} depends upon numerous parameters, the Figures 4.6–4.9 report only some possible situations that are very useful to frame the phenomena.

For example, the curves of Figure 4.6 (where $\xi_{sh} = 0$) show that the limit length d_2 of UGC are always very close to 10 km and that the constraint expressed by (4.36) is almost always decisive.

The curves of Figure 4.7 (computed for $X'' = 15\,\Omega$) show that (set $d_3 = 0$) the section UGC with $\xi_{sh} = 0.7$ can reach $d_2 = 40$ km if $d_1 = 0$ (being that the

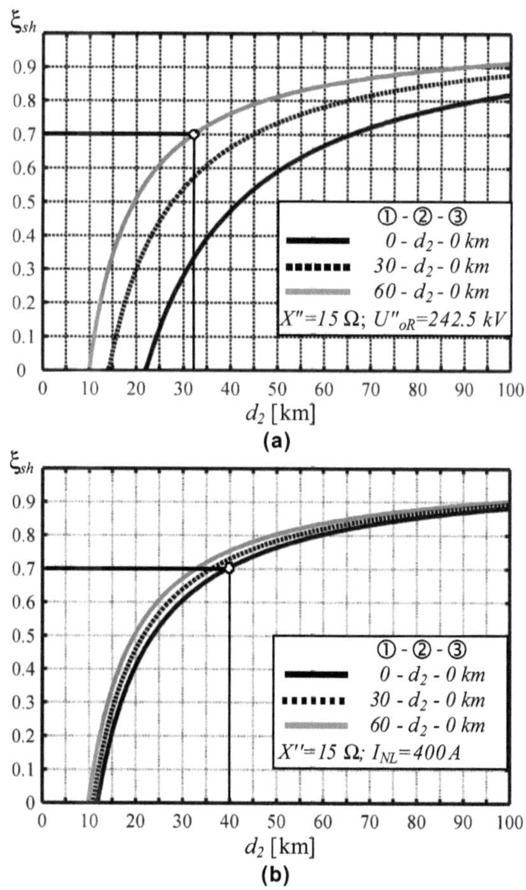

Figure 4.7 Examples of compensation degree ξ_{sh} as a function of UGC length d_2; (**a**) due to the constraint (4.35); (**b**) due to the constraint (4.36)

constraint (4.36) is decisive as shown in Figure 4.7b) or $d_2 = 32\,\text{km}$ if $d_1 = 60\,\text{km}$ (being that the constraint (4.35) is slightly decisive as shown in Figure 4.7a).

It is worth noting that the OHL ① prevailingly introduces a longitudinal reactance ($\cong 0.27\,\Omega/\text{km}$ at 50 Hz) which "increases" the short-circuit reactance X'' worsening the no-load energization subtransient regimes supplied by S.

Hence, it can be ascertained, by applying iteratively (4.33) and (4.34), that with $d_1 = 90\,\text{km}$ (by holding again $\xi_{sh} = 0.7$ and $d_3 = 0$) the constraint (4.35) would lessen the limit length to $d_2 = 26\,\text{km}$.

Figures 4.8 and 4.9 consider examples where $d_3 \neq 0$.

The overlapping of the three curves (grey, dotted, black) in Figures 4.8b and 4.9b is due to the particular lengths foreseen for d_1 and d_3 which always give rise to the same whole length of OHL (equal to 80 km) and change very little the no-load steady state with the UGC of length d_2.

Besides, Figures 4.8a and 4.9a highlight how the subtransient no-load regimes are affected by the three configurations in a different way.

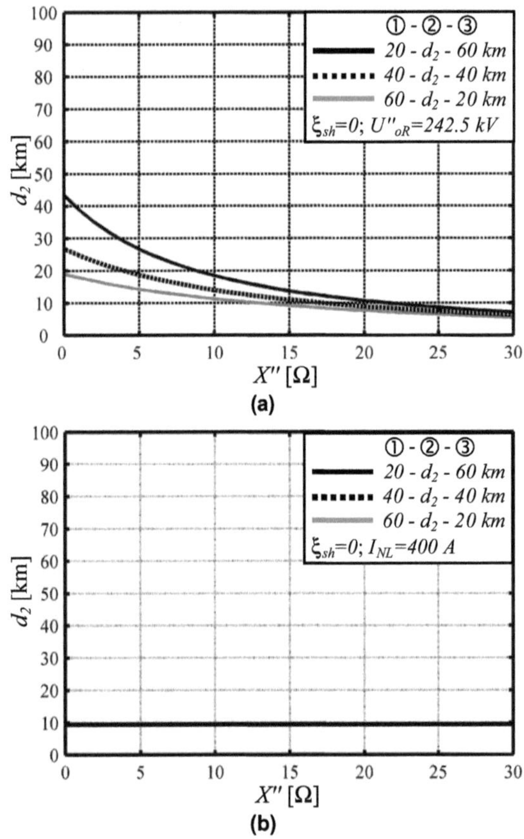

Figure 4.8 Examples of limit length d_2 for uncompensated UGC; (**a**) due to the constraint (4.35); (**b**) due to the constraint (4.36)

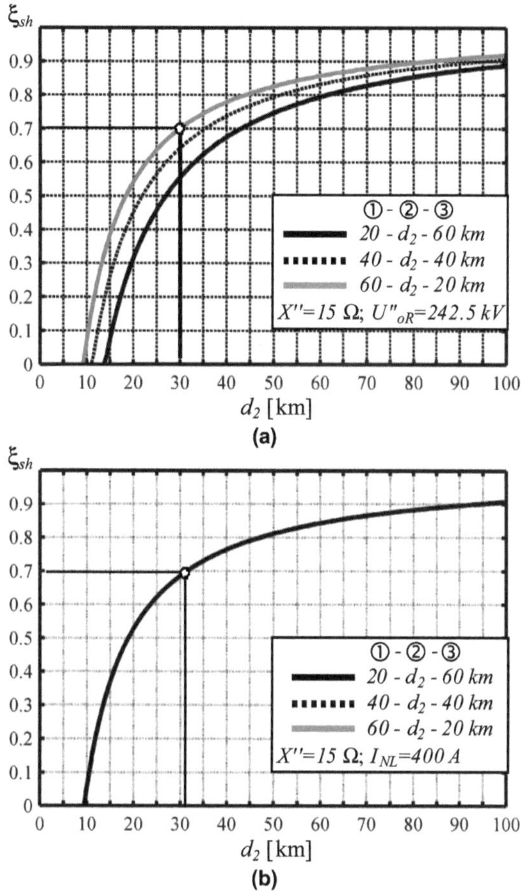

Figure 4.9 Examples of compensation degree ξ_{sh} as a function of UGC length d_2; (**a**) due to the constraint (4.35); (**b**) due to the constraint (4.36)

By hypothesising a shunt compensation of UGC with $\xi_{sh} = 0.7$, the length limit $d_2 = 32$ km can be derived graphically (in case of $d_1 = 20$ km and $d_3 = 60$ km), due to (Figure 4.9b) the constraint (4.36), whereas (in case of $d_1 = 60$ km and $d_3 = 20$ km) it is $d_2 = 30$ km due to (Figure 4.9a) the constraint (4.35).

For further example, by setting $d_1 = 90$ km and $d_3 = 20$ km, the limit length $d_2 = 21$ km, due to the constraint (4.35), can be derived numerically by applying (4.33) and (4.34).

It is evident that the energization of no-load mixed line from the node R must be evaluated by considering the lengths of three section in the right order: for example, if the order from S to R of the lengths d_1, d_2, d_3 is 60, 30, 20 km, when the energization from S is assessed the grey curve in Figure 4.9 must be considered, whereas the dark curve must be considered with energization from R.

If the equivalent generators at both ends were equal (same $U_0 = 230$ kV and same $X'' = 15\,\Omega$) it would be better the energization from R, which requires a lower compensation degree.

On the other hand, it appears very likely that the receiving-end is a very weak node (very high X'') so that it results less suitable than S for the energization of no-load line: also in these cases, (refer to Section 3.7) the detection of "best-end" switching suggested by CEGB [5] is advisable.

It is obvious that the present procedure is applicable to numerically analyse any situation without recurring to specific curves (e.g. change of the ports S and R, with magnitude $U_{0S} \neq 230\,\text{kV}$).

It is trivial to remember the possibility for the TSO of energising the line with voltage temporarily reduced.

These meaningful considerations on the energization and de-energization regimes must be followed by an in-depth analysis of power flows and network simulations in order to fix definitively the suitable compensation degree for UGC section.

It is worth remembering, as already stated in Section 3.7, that possible cases, where the steady state capacitive power ($Q_{\text{NL}} \approx -3U_0 I_{\text{NL}}$), absorbed by the no load mixed line, exceeds the ability of synchronous generators located in the neighbourhood (dangerous self excitation conditions), must be avoided.

As already stated, the hypothesis of uniformly distributed shunt compensation has been adopted: it has been throughout verified that both the capability charts and the line energization and de-energization phenomena remain almost unchanged by applying (with the same ξ_{sh}) lumped shunt compensations (see suitable matrices $\underline{\boldsymbol{M}}_{\xi}$ in Section 2.8.2) on length intervals not greater than 15–20 km.

Some attention must be paid (in the cases of lumped compensation) for the location of the limit I_c at H and K (in the matrix cascade).

4.8 The Use of Capability Charts as a Guide

The example of Figure 4.4 already gives an idea of the utility of the capability charts for transmission system operators.

It is worth remembering that the value of ξ_{sh} (evaluated after Section 4.7), is not only fundamental for a good line energization and de-energization but also determines satisfactory power flows well evident in the same figure.

Moreover the set of curves parameterized with phase voltages 220–240 kV highlights immediately how a good voltage level at R constitutes a further limiting constraint (chiefly for reactive power) within the capability charts for $I_c = 2800\,\text{A}$. Also the values of maximum transmissible active power $P_{R1} = 1850\,\text{MW}$ together with the power losses ΔP_1 and the compensation degree ξ_{sh} are very useful for economic evaluations: in particular, the compensation degree constitutes a direct reference to the installation cost of shunt reactors in section ②.

The no-load steady state current $I_{\text{NL}} = 391\,\text{A}$ and the corresponding reactive power $Q_{\text{NL}} = -270\,\text{Mvar}$ do not appear problematic.

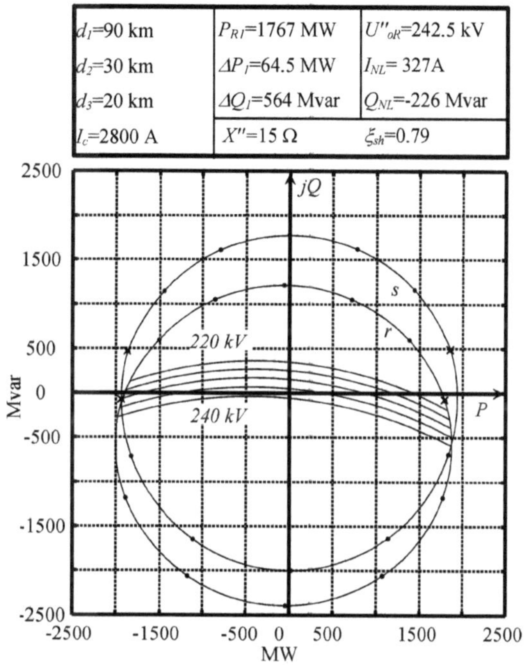

Figure 4.10 Second example of capability charts for mixed line

A second example (see Figure 4.10) highlights how the increased overall length of mixed line (90, 30, 20 km) yields now unacceptable voltage drops unless the node R has not capacitive load.

So it is possible to ascertain that when the mixed line length increases, the criterion of voltage drops and power losses becomes paramount: in correspondence of the power flow $P_{R1} = 1767$ MW it has $|\underline{U}_{0R}| = 210$ kV (as graphically inferable) unsatisfactory for network operation, a ratio $\Delta P_1 / P_{R1} = 3.65\%$ and also a reactive power $\Delta Q_1 = 564$ Mvar rather high from both technical and economical standpoints.

Figure 4.11 (third example) clearly shows how choosing a current limit of 2100 A (just below OHL warm month ampacity) gives regimes more rational than those of Figure 4.10.

Both Figures 4.12 and 4.13, which can represent some examples of mixed line for penetration into town centres ($d_3 = 0$), show how the decreased overall length, with reference to the previous case (110 km rather than 140 km), allows (in cold months) a higher chosen current limit (2500 A) and good voltage levels at R.

The comparison between Figures 4.14 and 4.12 underlines the reduction of capability charts due to the current limit (2100 A) and the subsequent better voltage levels at R.

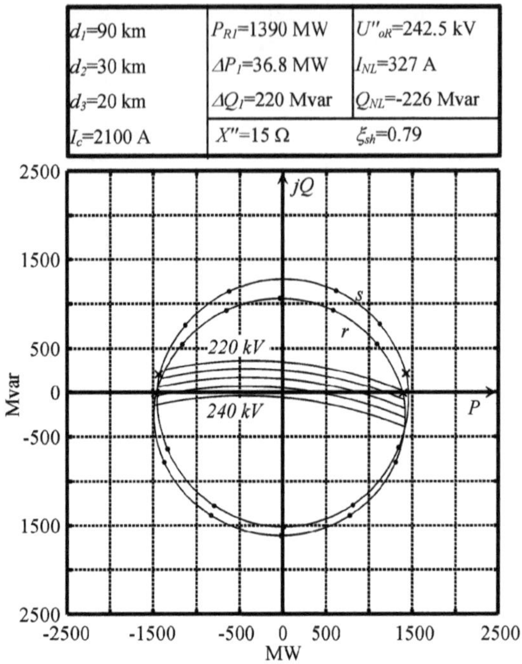

Figure 4.11 Third example of capability charts for mixed line

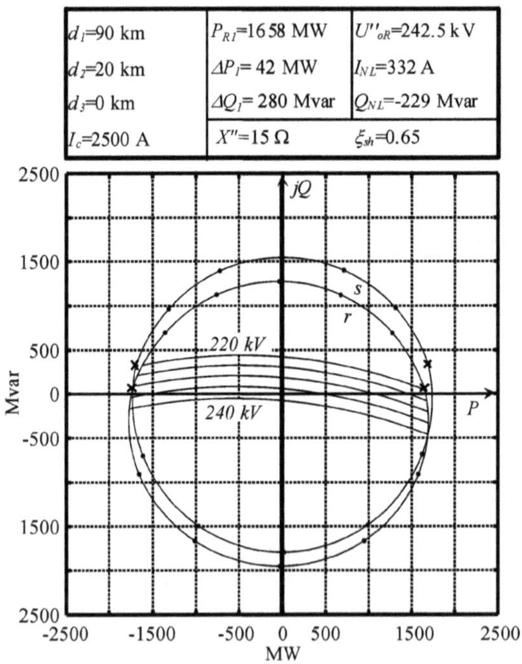

Figure 4.12 Fourth example of capability charts for mixed line

Figure 4.13 Fifth example of capability charts for mixed line

d_1=70 km	P_{R1}=1678 MW	U''_{oR}=242.5 kV
d_2=40 km	ΔP_1= 36 MW	I_{NL}=360 A
d_3=0 km	ΔQ_1= 189 Mvar	Q_{NL}=-249 Mvar
I_c=2500 A	X''=15 Ω	ξ_{sh}=0.79

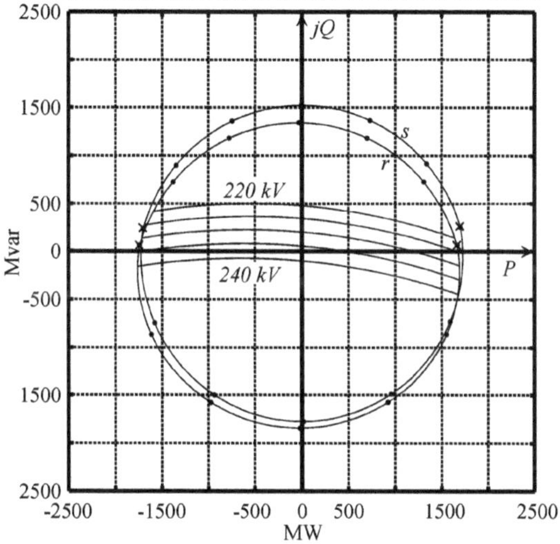

Figure 4.14 Sixth example of capability charts for mixed line

d_1=90 km	P_{R1}=1413 MW	U''_{oR}=242.5 kV
d_2=20 km	ΔP_1= 30 MW	I_{NL}=332 A
d_3=0 km	ΔQ_1=129 Mvar	Q_{NL}=-229 Mvar
I_c=2100 A	X''=15 Ω	ξ_{sh}=0.65

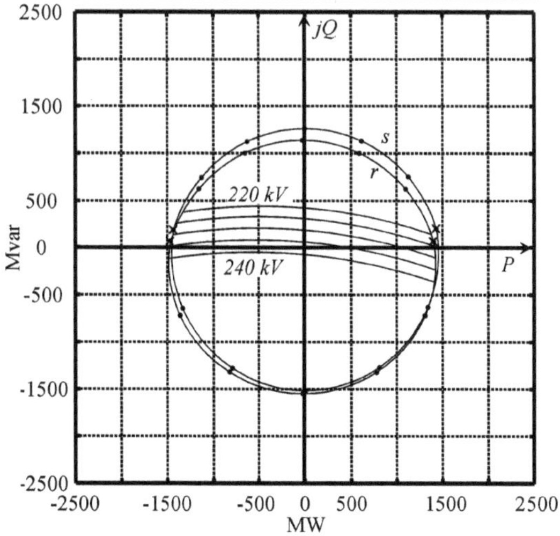

Figure 4.15 Example of capability charts of UGC line ($d_1 = d_3 = 0$); double-circuit with shunt compensation

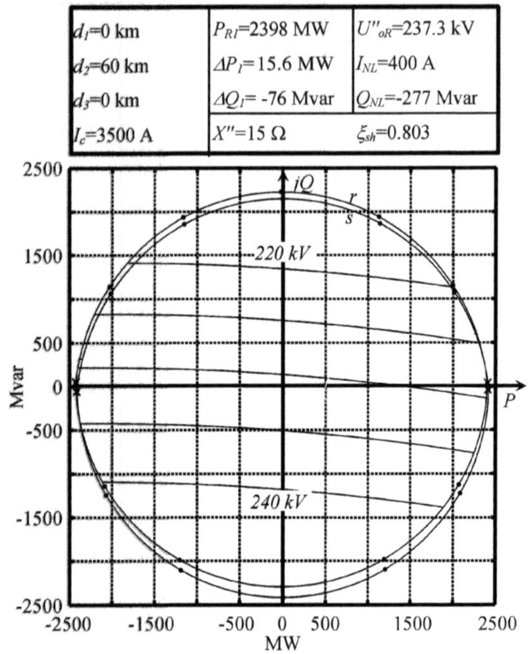

Figures 4.15 and 4.16 also appear very meaningful because they deal with cases where the line between S and R is entirely constituted by double-circuit UGC of 60 km (having set $d_1 = 0$, $d_2 = 60$, $d_3 = 0$).

Figure 4.15 shows the excellent performance ($P_{R1} = 2398$ MW) of 60 km compensated double-circuit cable (due to its very high ampacity limit 3500 A) with good voltage levels because of its low longitudinal impedance.

It is worth noting that the corresponding technical-economical burden in this case is rather high since the overall shunt reactive compensation power is (at 50 Hz) $Q_\xi = 0.804 \cdot (23\,\text{Mvar/km}) \cdot (60\,\text{km}) = 1100$ Mvar and must be shared at least in three sections along the line (see Section 2.8.2).

The prohibitive conditions which would exist in the case of uncompensated UGC (with the same length) are shown in Figure 4.16: the highest steady state no-load reactive power $Q_{\text{NL}} = -1423$ Mvar with corresponding $I_{\text{NL}} > 2000$ A and the subtransient voltage level $U''_{0R} = 272$ kV during energization.

It is also of note that the typical shape of capability charts with cusps at points **x** recurs chiefly each time there is a low compensated UGC as in Chapter 3.

In these cases (chiefly with high d_2 and low ξ_{sh}), the differences $\underline{S}_S - \underline{S}_R$ appear graphically very evident due to the strong reactive (capacitive) power absorbed by the cable.

In the mixed lines (chiefly with high d_1, d_3 and ξ_{sh}) the reactive power absorbed by the overall link becomes of inductive nature (see Figures 4.4 and 4.10–4.14), being prevalent the inductive demand of OHL.

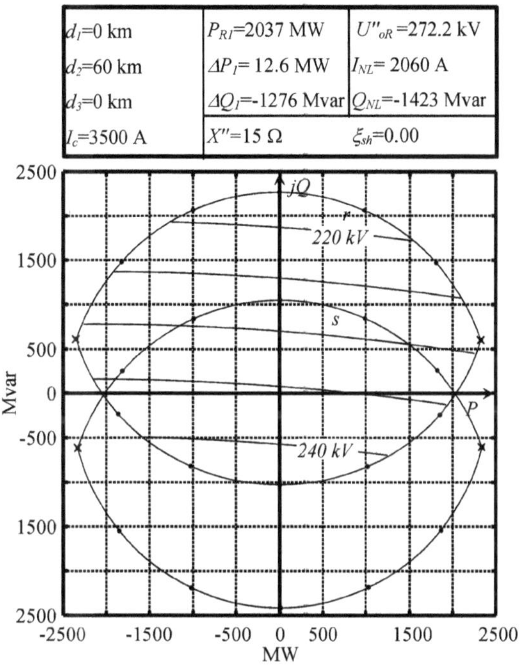

Figure 4.16 The prohibitive conditions ($\underline{U''_{0R}}$; I_{NL}; ΔQ_1) of uncompensated double-circuit UGC (60 km)

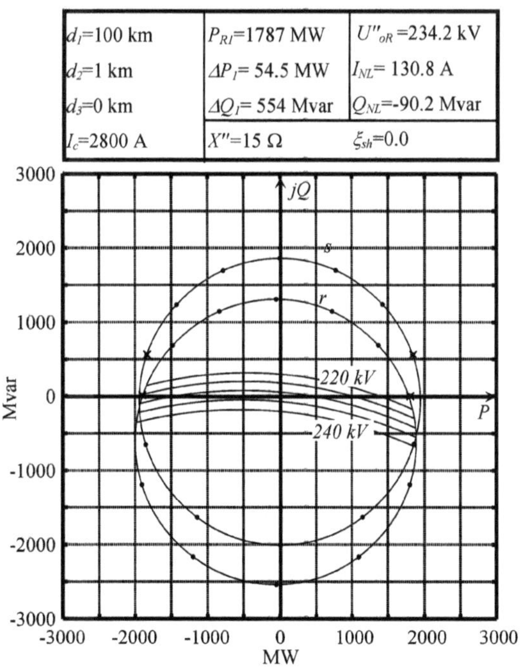

Figure 4.17 Example of capability charts for 100 km OHL and 1 km UGC

The proposed procedure has been extensively applied to many mixed line configurations but can give extremely useful suggestions for entirely overhead line as well: if, in the configuration of Figure 4.1, the length of UGC section is fixed as $d_2 \ll d_1 + d_3$, it is possible to highlight with good accuracy the foreseeable behaviour of the considered OHL.

A meaningful example is given by the capability charts of Figure 4.17 which demonstrates very restricted limits for satisfying voltage operations (owing to high voltage drops), high active losses (see 54.5 MW!) and a huge quantity of inductive power absorbed (see 554 Mvar!). This kind of operation (at the thermal limit) is often due to the big hindrances in constructing new OHLs, in several countries, which compel to make maximum use of the existing overhead grid. Alternatively, the procedures of Chapter 3 can be suitably used by introducing the "positive sequence model" of OHL (instead of UGC) with $d = 100\,\text{km}$.

4.9 "Receiving Area" and "Sending Area" as Intersections of Sets

Once imposed (as in 1^ analysis) the phasors $I_K \angle 0 \equiv I_c$ and $\underline{U}_{0S}(\delta)$, if (4.28) is made explicit it is possible, being

$$\underline{N}_{R1} = \begin{array}{|c|c|} \hline \underline{A}_{R1} & \underline{B}_{R1} \\ \hline \underline{C}_{R1} & \underline{D}_{R1} \\ \hline \end{array}\,, \qquad \begin{aligned} \underline{A}_{R1} &= \underline{D}_{123} - \underline{B}_{123}\underline{C}_{12}/\underline{A}_{12}\,, \\ \underline{B}_{R1} &= -\underline{B}_{123}/\underline{A}_{12}\,, \\ \underline{C}_{R1} &= -\underline{C}_{123} + \underline{A}_{123}\underline{C}_{12}/\underline{A}_{12}\,, \\ \underline{D}_{R1} &= \underline{A}_{123}/\underline{A}_{12}\,, \end{aligned}$$

to obtain (4.37)

$$\begin{aligned} \frac{\underline{S}_R(1\wedge)}{3} &= [\underline{A}_{R1}\underline{U}_{0S} + \underline{B}_{R1} I_K] \cdot [\underline{C}_{R1}\underline{U}_{0S} + \underline{D}_{R1} I_K]^* \\ &= \underline{A}_{R1}\underline{C}^*_{R1}|\underline{U}_{0S}|^2 + \underline{B}_{R1}\underline{D}^*_{R1} I_K^2 \\ &\quad + \underline{A}_{R1}\underline{D}^*_{R1} I_K \underline{U}_{0S}(\delta) + \underline{B}_{R1}\underline{C}^*_{R1} I_K \underline{U}^*_{0S}(\delta)\,. \end{aligned} \tag{4.37}$$

Equation 4.37 gives the values of $\frac{\underline{S}_R(1\wedge)}{3}$ for $\delta = 0$–2π, without respecting the limit I_c at node H.

Analogously once imposed (as in 2^ analysis) the phasors $I_H \angle 0 \equiv I_c$ and $\underline{U}_{0S}(\vartheta)$, if (4.29) is made explicit it is possible, being that

$$\underline{N}_{R2} = \begin{array}{|c|c|} \hline \underline{A}_{R2} & \underline{B}_{R2} \\ \hline \underline{C}_{R2} & \underline{D}_{R2} \\ \hline \end{array}\,, \qquad \begin{aligned} \underline{A}_{R2} &= \underline{D}_{123} - \underline{B}_{123}\underline{C}_{1}/\underline{A}_{1}\,, \\ \underline{B}_{R2} &= -\underline{B}_{123}/\underline{A}_{1}\,, \\ \underline{C}_{R2} &= -\underline{C}_{123} + \underline{A}_{123}\underline{C}_{1}/\underline{A}_{1}\,, \\ \underline{D}_{R2} &= \underline{A}_{123}/\underline{A}_{1}\,, \end{aligned}$$

to obtain (4.38)

$$\begin{aligned}\frac{\underline{S}_R(2^\wedge)}{3} &= [\underline{A}_{R2}\underline{U}_{0S} + \underline{B}_{R2}\underline{I}_H]\cdot[\underline{C}_{R2}\underline{U}_{0S} + \underline{D}_{R2}\underline{I}_H]^* \\ &= \underline{A}_{R2}\underline{C}^*_{R2}|\underline{U}_{0S}|^2 + \underline{B}_{R2}\underline{D}^*_{R2}I^2_H \\ &\quad + \underline{A}_{R2}\underline{D}^*_{R2}I_H\underline{U}_{0S}(\vartheta) + \underline{B}_{R2}\underline{C}^*_{R2}I_H\underline{U}^*_{0S}(\vartheta)\,. \end{aligned} \tag{4.38}$$

Equation 4.38 gives the values of $\frac{\underline{S}_R(2^\wedge)}{3}$ for $\vartheta = 0\text{–}2\pi$, without respecting for the limit I_c at node K.

The simultaneous representation of both curves $\underline{S}_R(1^\wedge)$ and $\underline{S}_R(2^\wedge)$ which bound each other, allows drawing (once fixed I_c) the bound r of "receiving-end power area" (see Figures 4.18 and 4.19).

With similar procedure, by developing (4.8) and (4.20), it is possible to have (4.39) and (4.40)

$$\frac{\underline{S}_S(1^\wedge)}{3} = \underline{U}_{0S}[\underline{C}_{S1}\underline{U}_{0S} + \underline{D}_{S1}\underline{I}_K]^* = \underline{C}^*_{S1}|\underline{U}_{0S}|^2 + \underline{I}_K\underline{D}^*_{S1}\underline{U}_{0S}(\delta)\,, \tag{4.39}$$

$$\frac{\underline{S}_S(2^\wedge)}{3} = \underline{U}_{0S}[\underline{C}_{S2}\underline{U}_{0S} + \underline{D}_{S2}\underline{I}_H]^* = \underline{C}^*_{S2}|\underline{U}_{0S}|^2 + \underline{I}_H\underline{D}^*_{S2}\underline{U}_{0S}(\vartheta)\,, \tag{4.40}$$

so that, by means of $\underline{S}_S(1^\wedge)$ and $\underline{S}_S(2^\wedge)$, the bound s of "sending-end power area" can be drawn. Equations 4.39 and 4.40 give circumferences, whereas (4.37) and (4.38) give ellipses with low eccentricity since beyond the third term with $\underline{U}_{0S}$ there is a forth term (generally slight) with the complex conjugate $\underline{U}^*_{0S}$.

This kind of approach, which determines the sending-end power area and receiving-end power area as *intersection of sets*, can become very useful to draw the

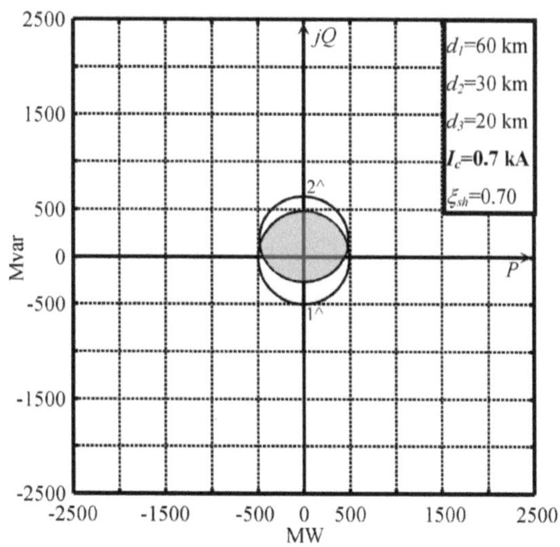

Figure 4.18 Delimiting the receiving areas by means of (4.37) (curve 1^) and (4.38) (curve 2^)

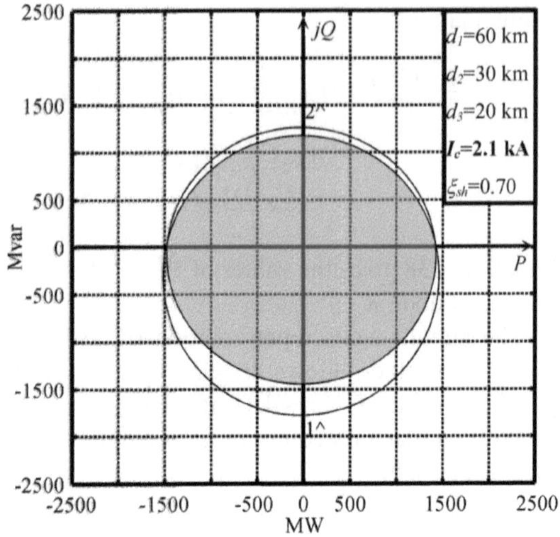

Figure 4.19 Delimiting the receiving areas by means of (4.37) (curve 1^) and (4.38) (curve 2^)

capability charts of a line with many and different (for location and magnitude) ampacity constraints.

4.10 Analysis Completion

It can be useful and suitable to complete the analysis of the system in order to know and qualify in detail different regimes concerning complex power more or less far from the capability chart area outlines, both those located inside the capability area (surely compatible with the current constraints in (4.1) and (4.2)), and those located outside (which surely give more or less strong derogations of the same constraints); in some cases, it is furthermore interesting to verify how the constraints themselves are respected along the line.

For these investigations, it is extremely useful to take advantage of (analogously to Chapter 3) Ossanna's method and the matrix algorithms.

4.10.1 Analysis Completion by Means of Ossanna's Method and Matrix Algorithms

Before proceeding with the use of Ossanna's formulae reported in Section 3.9.2 it is necessary and sufficient to explicitly state that in these cases of mixed lines $\underline{A} \equiv \underline{A}_{123}$ and $\underline{B} \equiv \underline{B}_{123}$ must be always set.

In this way, once fixed, any complex power inside the receiving-end power area, one immediately obtains the corresponding voltage and current phasors at R, which allows determining not only the corresponding regime at sending-end port (so that

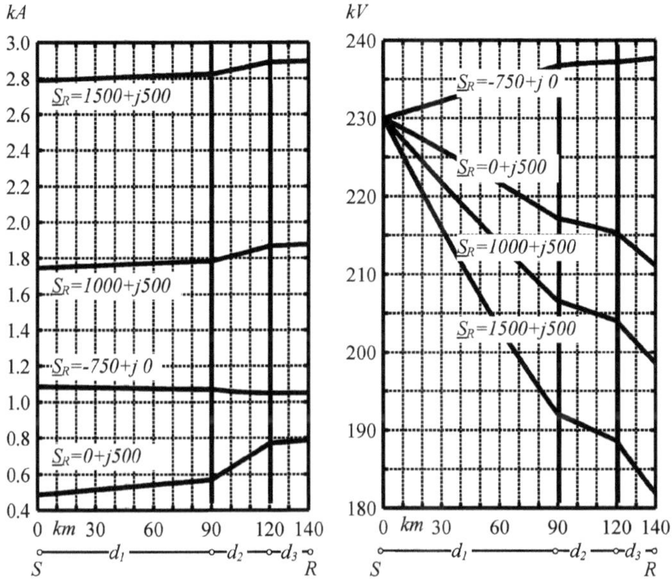

Figure 4.20 Current and voltage magnitudes along the mixed line (see Figure 4.10; $d_1 = 90\,\text{km}$; $d_2 = 30\,\text{km}$; $d_3 = 20\,\text{km}$; $\xi_{sh} = 0.79$)

the transmission angle and power losses can be immediately computed), but also (by means of suitable matrix algorithms detailed in the Chapter 2) the regimes at different locations along the line.

It is interesting to highlight how the method can be used also to evaluate and locate the stresses in the systems during possible low differences in the power outside the capability outlines.

Figure 4.20 shows an immediate application of Ossanna's method: the current and voltage magnitudes along the mixed line of Figure 4.10 are clearly detectable for the four shown complex power required at receiving-end. It is worth noting that $\underline{S}_R = 1500\,\text{MW} + \text{j}\,500\,\text{Mvar}$ lies slightly outside the receiving capability area of Figure 4.10: in fact, the current magnitudes of Figure 4.20 exceed, slightly along the line, the ampacity of 2800 A up to 2900 A at port R (with too high voltage drops).

4.11 Circuital Considerations

It is worth premising that the considerations which will be developed in this paragraph give a useful contribution for the direct interpretation of the matrix relations obtained up to here, allow giving the single elements of the matrices themselves simpler expressions (even if with exact equivalence): the circuital analysis could represent a useful exercise for a student of electrical engineering.

4.11.1 The Three Matrices $\underline{N}_{H1}$, $\underline{N}_{S1}$, $\underline{N}_{R1}$

In the following, the matrix relations which summarize in symbolic form the above presented equations (4.11), (4.8) and (4.28) are reported:

$$\begin{bmatrix} \underline{U}_{0H} \\ \underline{I}_H \end{bmatrix} = \underbrace{\begin{bmatrix} \underline{A}_{H1} & \underline{B}_{H1} \\ \underline{C}_{H1} & \underline{D}_{H1} \end{bmatrix}}_{\underline{N}_{H1}} \begin{bmatrix} \underline{U}_{0S} \\ \underline{I}_K \end{bmatrix}, \tag{4.11}$$

$$\begin{bmatrix} \underline{U}_{0S} \\ \underline{I}_S \end{bmatrix} = \underbrace{\begin{bmatrix} 1 & 0 \\ \underline{C}_{S1} & \underline{D}_{S1} \end{bmatrix}}_{\underline{N}_{S1}} \begin{bmatrix} \underline{U}_{0S} \\ \underline{I}_K \end{bmatrix}, \tag{4.8}$$

$$\begin{bmatrix} \underline{U}_{0R} \\ \underline{I}_R \end{bmatrix} = \underbrace{\begin{bmatrix} \underline{A}_{R1} & \underline{B}_{R1} \\ \underline{C}_{R1} & \underline{D}_{R1} \end{bmatrix}}_{\underline{N}_{R1}} \begin{bmatrix} \underline{U}_{0S} \\ \underline{I}_K \end{bmatrix}, \tag{4.28}$$

They are suitable for meaningful interpretations of circuital nature and for the simple determination of the parameters, as it will be presented in the Sections 4.11.2–4.11.4.

4.11.2 The Elements of $\underline{N}_{H1}$

By expanding (4.11), it yields

$$\underline{U}_{0H} = \underline{A}_{H1}\underline{U}_{0S} + \underline{B}_{H1}\underline{I}_K \,, \tag{4.41}$$

$$\underline{I}_H = \underline{C}_{H1}\underline{U}_{0S} + \underline{D}_{H1}\underline{I}_K \,, \tag{4.42}$$

where for the elements $\underline{A}_{H1}$, $\underline{B}_{H1}$, $\underline{C}_{H1}$, $\underline{D}_{H1}$ precise expressions have already been obtained in Section 4.4.

Now it is possible to deduce new equivalent expressions for the same elements. In fact, the first addendum of (4.41) i.e. $\underline{A}_{H1}\underline{U}_{0S} = \underline{U}^{\bullet}_{0H}$ represents the voltage at H due to the application of the voltage phasor $\underline{U}_{0S}$ when the current I_K at K is zero (see Figure 4.21).

In this perspective, it is important to note that the regime with $I = 0$ at K is not in contrast with the metallic continuity in K itself, since the port R is supposed to be also active. So by adopting the symbols shown in Figure 4.21, it results immediate (on the basis of the fundamental transmission equations as in Chapter 2) to write (being that $I_K = 0$) also

$$\underline{U}_{0S} = \underline{A}_{12}\underline{U}^{\bullet}_{0K} \,; \quad \underline{U}^{\bullet}_{0H} = \underline{A}_{2}\underline{U}^{\bullet}_{0K} \tag{4.43}$$

from which it yields

$$\underline{U}^{\bullet}_{0H} = (A_2/A_{12})\underline{U}_{0S} \quad \Rightarrow \quad \underline{A}_{H1} = \underline{A}_2/\underline{A}_{12}\,. \tag{4.44}$$

For the determination of unknown $\underline{C}_{H1}$, the first addendum of (4.42) i.e. $\underline{C}_{H1}\underline{U}_{0S} = \underline{I}^{\bullet}_{H}$ can be interpreted as the current at H due to the application of the voltage phasor $\underline{U}_{0S}$ when $\underline{I} = 0$ is imposed at K so that:

$$\underline{I}^{\bullet}_{H} = \underline{C}_2\underline{U}^{\bullet}_{0K} \quad \text{and} \quad \underline{U}^{\bullet}_{0K} = \frac{\underline{U}_{0S}}{\underline{A}_{12}}\,. \tag{4.45}$$

It is immediate to obtain

$$\underline{I}^{\bullet}_{H} = \underline{C}_2\underline{U}^{\bullet}_{0K} = \frac{\underline{C}_2}{\underline{A}_{12}} \cdot \underline{U}_{0S} \quad \Rightarrow \quad \underline{C}_{H1} = \frac{\underline{C}_2}{\underline{A}_{12}} \tag{4.46}$$

By following similar considerations and passages (with reference to the scheme and symbols of Figure 4.22), it is possible to derive expressions in order to compute also the elements $\underline{B}_{H1}$ and $\underline{D}_{H1}$:

$$\underline{U}^{\bullet\bullet}_{0H} = -\underline{B}_1\underline{I}^{\bullet\bullet}_{S}\,; \quad \underline{I}_K = \underline{A}_{12} \cdot \underline{I}^{\bullet\bullet}_{S}\,; \quad \underline{U}^{\bullet\bullet}_{0H} = -\frac{\underline{B}_1}{\underline{A}_{12}}\underline{I}_K \quad \Rightarrow \quad \underline{B}_{H1} = -\frac{\underline{B}_1}{\underline{A}_{12}}\,; \tag{4.47}$$

$$\underline{I}_K = \underline{A}_{12}\underline{I}^{\bullet\bullet}_{S}\,; \quad \underline{I}^{\bullet\bullet}_{H} = \underline{A}_1 \cdot \underline{I}^{\bullet\bullet}_{S}\,; \quad \underline{I}^{\bullet\bullet}_{H} = \frac{\underline{A}_1}{\underline{A}_{12}}\underline{I}_K \quad \Rightarrow \quad \underline{D}_{H1} = \frac{\underline{A}_1}{\underline{A}_{12}}\,. \tag{4.48}$$

The reader can ascertain that the four elements of $\underline{N}_{H1}$ computed by (4.44)–(4.48) agree perfectly with those computed by (4.11) already presented in Section 4.4, so that the phasorial diagram of Figure 4.2 and the references in the same figure hold their clear meaning.

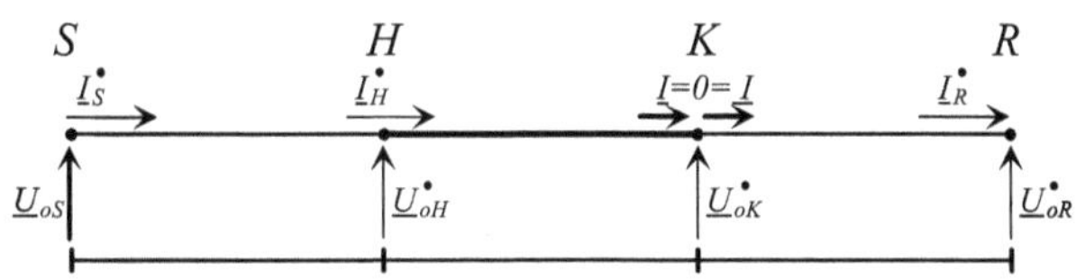

Figure 4.21 Scheme for the determination of $\underline{A}_{H1}$ and $\underline{C}_{H1}$

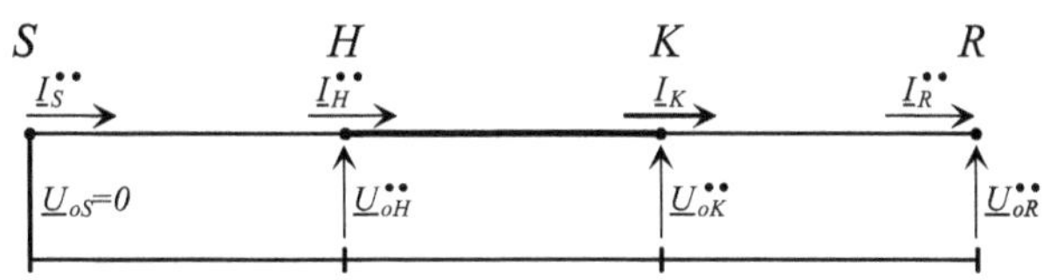

Figure 4.22 Scheme for the determination of $\underline{B}_{H1}$ and $\underline{D}_{H1}$

In particular, it can be stated in general terms that

$$\eta_1 = \text{angle}(\underline{C}_{H1}) \,, \tag{4.49}$$
$$\mu_1 = \text{angle}(\underline{D}_{H1}\underline{I}_K) \,, \tag{4.50}$$

and that the phasor $\underline{U}_{0S}(\delta)$ determines, in the interval δ_1–δ_2, current phasor $\underline{I}_H$ within the limit $\underline{I}_c$, being obviously that

$$\delta_1 = \pi - \beta_1 - \eta_1 + \mu_1 \,, \tag{4.51}$$
$$\delta_2 = \delta_1 + \beta_1 \,. \tag{4.52}$$

4.11.3 The Matrix $\underline{N}_{S1}$

The two elements of the matrix $\underline{N}_{S1}$ can be derived by applying the criteria already presented in the previous paragraphs: the fundamental expressions and the successive passages do not need explanations but noting that (4.53) regards Figure 4.21 and (4.54) Figure 4.22.

$$\underline{U}_{0S} = \underline{A}_{12}\underline{U}^{\bullet}_{0K} \,; \quad \underline{I}^{\bullet}_S = \underline{C}_{12} \cdot \underline{U}^{\bullet}_{0K} \,; \quad \underline{I}^{\bullet}_S = \frac{\underline{C}_{12}}{\underline{A}_{12}}\underline{U}_{0S} \quad \Rightarrow \quad \underline{C}_{S1} = \frac{\underline{C}_{12}}{\underline{A}_{12}} \,; \tag{4.53}$$

$$\underline{I}_K = \underline{A}_{12}\underline{I}^{\bullet\bullet}_S \,; \qquad \underline{I}^{\bullet\bullet}_S = \frac{1}{\underline{A}_{12}}\underline{I}_K \quad \Rightarrow \quad \underline{D}_{S1} = \frac{1}{\underline{A}_{12}} \,. \tag{4.54}$$

4.11.4 The Matrix $\underline{N}_{R1}$

In the following the elements of the matrix $\underline{N}_{R1}$ are reported, by remembering that (4.55) and (4.56) refer to Figure 4.21 whereas (4.57) and (4.58) to Figure 4.22.

$$\underline{U}_{0S} = \underline{A}_{12}\underline{U}^{\bullet}_{0K} \,; \quad \underline{U}^{\bullet}_{0R} = \underline{D}_3 \cdot \underline{U}^{\bullet}_{0K} \,;$$
$$\underline{U}^{\bullet}_{0R} = \frac{\underline{D}_3}{\underline{A}_{12}}\underline{U}_{0S} \quad \Rightarrow \quad \underline{A}_{R1} = \frac{\underline{D}_3}{\underline{A}_{12}} \,; \tag{4.55}$$

$$\underline{U}_{0S} = \underline{A}_{12}\underline{U}^{\bullet}_{0K} \,; \quad \underline{I}^{\bullet}_R = -\underline{C}_3 \cdot \underline{U}^{\bullet}_{0K} \,;$$
$$\underline{I}^{\bullet}_R = -\frac{\underline{C}_3}{\underline{A}_{12}}\underline{U}_{0S} \quad \Rightarrow \quad \underline{C}_{R1} = -\frac{\underline{C}_3}{\underline{A}_{12}} \,; \tag{4.56}$$

$$\underline{I}_K = \underline{A}_{12} \cdot \underline{I}^{\bullet\bullet}_S \,; \quad U^{\bullet\bullet}_{0R} = -\underline{B}_{123}\underline{I}^{\bullet\bullet}_S \,;$$
$$U^{\bullet\bullet}_{0R} = -\frac{\underline{B}_{123}}{\underline{A}_{12}}\underline{I}_K \quad \Rightarrow \quad \underline{B}_{R1} = -\frac{\underline{B}_{123}}{\underline{A}_{12}} \,; \tag{4.57}$$

$$\underline{I}_R^{\bullet\bullet} = \underline{A}_{123}\underline{I}_S^{\bullet\bullet}\,; \qquad \underline{I}_K = \underline{A}_{12}\cdot\underline{I}_S^{\bullet\bullet}\,;$$

$$\underline{I}_R^{\bullet\bullet} = \frac{\underline{A}_{123}}{\underline{A}_{12}}\underline{I}_K \qquad \Rightarrow \qquad \underline{D}_{R1} = \frac{\underline{A}_{123}}{\underline{A}_{12}}\,. \tag{4.58}$$

4.11.5 The Matrices $\underline{N}_{K2}$, $\underline{N}_{S2}$, $\underline{N}_{R2}$

In the following, the matrix relations which summarize in symbolic form the above presented equations (4.23), (4.20) and (4.29) are reported:

$$\begin{bmatrix} \underline{U}_{0K} \\ \underline{I}_K \end{bmatrix} = \underbrace{\begin{bmatrix} \underline{A}_{K2} & \underline{B}_{K2} \\ \underline{C}_{K2} & \underline{D}_{K2} \end{bmatrix}}_{\underline{N}_{K2}} \begin{bmatrix} \underline{U}_{0S} \\ \underline{I}_H \end{bmatrix}, \tag{4.23}$$

$$\begin{bmatrix} \underline{U}_{0S} \\ \underline{I}_S \end{bmatrix} = \underbrace{\begin{bmatrix} 1 & 0 \\ \underline{C}_{S2} & \underline{D}_{S2} \end{bmatrix}}_{\underline{N}_{S2}} \begin{bmatrix} \underline{U}_{0S} \\ \underline{I}_H \end{bmatrix}, \tag{4.20}$$

$$\begin{bmatrix} \underline{U}_{0R} \\ \underline{I}_R \end{bmatrix} = \underbrace{\begin{bmatrix} \underline{A}_{R2} & \underline{B}_{R2} \\ \underline{C}_{R2} & \underline{D}_{R2} \end{bmatrix}}_{\underline{N}_{R2}} \begin{bmatrix} \underline{U}_{0S} \\ \underline{I}_H \end{bmatrix}, \tag{4.29}$$

Also for the elements of these matrices, the simple expressions hereafter reported can be obtained (by following the same methodology of circuital analysis). However, they have full agreement (via numerical validation) with those previously derived by matrix algorithms.

$$\underline{N}_{K2}: \quad \underline{A}_{K2} = \frac{\underline{D}_2}{\underline{A}_1}\,; \quad \underline{B}_{K2} = -\frac{\underline{B}_{12}}{\underline{A}_1}\,; \quad \underline{C}_{K2} = -\frac{\underline{C}_2}{\underline{A}_1}\,; \quad \underline{D}_{K2} = \frac{\underline{A}_{12}}{\underline{A}_1}\,; \tag{4.59}$$

$$\underline{N}_{S2}: \quad \underline{C}_{S2} = \frac{\underline{C}_1}{\underline{A}_1}\,; \quad \underline{D}_{S2} = \frac{1}{\underline{A}_1}\,; \tag{4.60}$$

$$\underline{N}_{R2}: \quad \underline{A}_{R2} = \frac{\underline{D}_{23}}{\underline{A}_1}\,; \quad \underline{B}_{R2} = -\frac{\underline{B}_{123}}{\underline{A}_1}\,; \quad \underline{C}_{R2} = -\frac{\underline{C}_{23}}{\underline{A}_1}\,; \quad \underline{D}_{R2} = \frac{\underline{A}_{123}}{\underline{A}_1}\,. \tag{4.61}$$

4.12 Conclusions

The cable lines together with mixed ones can lessen sensibly many route problems by virtue of their adaptability in the territory and have become a recurring topic in the technical literature [6].

The authors have already presented comprehensive algorithms to deeply analyse the transmission performances of UGC in Chapter 3.

In this chapter, the method has been further developed and extended to take into account mixed lines composed of overhead and cable lines.

Also in these cases, the capability charts can be created, which highlights at quick glance the whole field of transmissible complex power compatibly with the assumed constraints for both current and voltage and are enhanced with other power transmission features, which result very useful to evaluate the mixed link operation.

This gives a meaningful contribution to both the grid planning and operation.

The cpu-time to compute a capability chart is about four seconds (PC Pentium 2.8 GHz, RAM 1048 MB). The procedure is versatile as well since it can be easily applied to more simple mixed lines ($d_1 = 0$ or $d_3 = 0$), to only cable line ($d_1 = 0 = d_3$) and to other configurations (e.g. Gas Insulated Lines).

This chapter is also a powerful guide to choose the shunt compensation degree of UGC in order to take under control the transients and steady state regimes of energization and de-energization.

Although the hypothesis of uniformly distributed compensation has been adopted, the method is easily applicable to lumped one (at each ends or also at intermediate locations).

References

1. R. Benato, A. Paolucci: Operating capability of ac EHV mixed lines with overhead and cables links, *Electric Power System Research*, Vol. 78/4, April 2008, pp. 584–594
2. R. Benato, A. Paolucci: Power flow in multiconductor asymmetrical systems. A PC program, *L'Energia Elettrica*, Vol. 77, No. 6, November–December 2000, pp. 43–53 (in Italian)
3. R. Benato: Multiconductor Cell Analysis Of Power Cable Steady-State, *Proc. of JICABLE 2007, Paper N C.5.2.9.*, pp. 707–712, Versailles, France, 24–28 June
4. R. Benato: Multiconductor Analysis of Underground Power Transmission Systems: EHV AC Cables, *Electric Power System Research*, Vol. 79, Issue 1, January 2009, pp. 27–38
5. British Electricity International, Modern Power Station Practice, EHV Transmission, Vol. K, Chapter 9, Sect. 3.3: Control of switching surges, Pergamon Press (Third Edition, 1991, Oxford)
6. F.M. Gatta, S. Lauria: Very long EHV cables and mixed overhead-cable lines. Steady-state operation, *Proc. of IEEE St. Petersburg Power Tech'05 Conference*, St. Petersburg, Russia, 2005

Chapter 5
Multiconductor Analysis of UGC

5.1 Introduction

The cable systems constitute paradigmatic and complex cases of multiconductor systems which cannot be studied in detail by means of a simplified single phase equivalent circuit.

In the electrical system research, the multiconductor theory is used in several situations. A great attention has always been devoted to the analysis of electromagnetic interferences between systems of different kind, e.g. in order to evaluate possible harmful effects due to AC power lines on telecommunication lines or metallic pipes: it has been well treated in a wide and detailed scientific literature [1–3].

Obviously, the modelling used for the above mentioned electromagnetic compatibility studies can be applied to the multiconductor analysis inside a unique power system [4, 5] where besides the overhead lines there are also cable lines with their metallic sheaths (UGC) and gas insulated lines with their metallic enclosures (GIL).

On the other hand, the usual overhead lines at power frequency would also require a multiconductor approach if conductors such as the earth wires and the counterpoises are realistically considered and the configuration asymmetry is not neglected.

Moreover, in any system the soil is not a filiform conductor.

Hence it is easy to understand how, considering the physical reality of the power networks, it can be questionable to assume purely three-phase configurations and perfectly symmetrical ones, so to use the three sequence modelling; in many cases, the multiconductor analysis becomes necessary, since it allows one to achieve (with the suitable procedures) great precision results so offering a powerful tool in order to validate (or less) approximated and simplified computation methods.

It is worth premising that in this chapter examples of the multiconductor procedures will be applied to a case already considered in Chapter 3 (i.e. cable #b, in cross-bonding with phase transpositions, $d = 60\,\text{km}$) and in particular to:

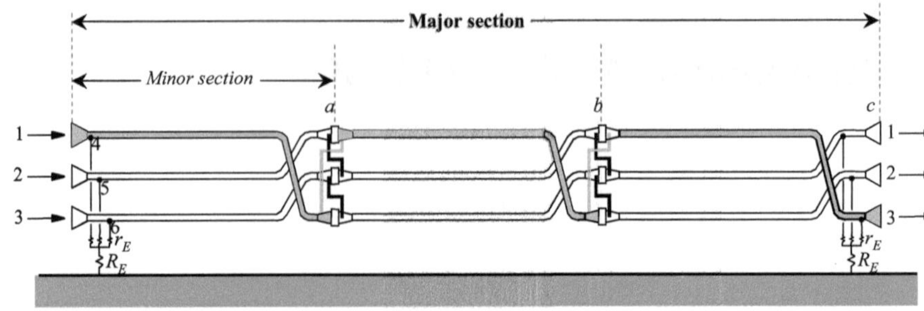

Figure 5.1 Cross-bonded single-circuit cable with phase transpositions

- two steady state regimes (i.e. points 1 and 6 of Figure 3.36) of the capability charts ($U_{0S} = 230\,\text{kV}$) with lumped shunt compensation every 20 km, also checking not to exceed the ampacity;
- subtransient no-load energization with and without lumped shunt compensation.

In the development of the procedures, the multiconductor analysis will individuate different circuital models, each characterized by their own admittance matrix: the suitable final composition of these partial matrices will yield the overall matrix which completely characterizes the system and gives the solution algorithms.

In particular the modelling of 60 km long UGC deals with the cascade connection of 32 "major sections" (see Figure 5.1) where each is cross-bonded in order to avoid the circulation of induced currents in the sheaths, in spite of the bonding (r_E) and earthing connections (R_E) at the ends.

5.2 Multiconductor Cell of Three Single-Core Cables Lines

In order to gradually develop an exposition of the criteria for the modelling and the analysis of the multiconductor system, let us consider (see Figure 5.2) a stretch of length $\Delta_\ell \leq 0.3$ km between the two sections S and R of six conductors (three phases and three sheaths) parallel to themselves and to the ground surface where earth return current flows; if $d \gg \Delta_\ell$ is assumed, the border effects can be neglected.

In such a case, the treatment (given by Carson [6], Pollaczek [7] and effectively expounded and noted in [8]), shows (if the transversal couplings due to the phase-to-sheath and sheath-to-earth conductive-capacitive susceptances are treated separately) how the longitudinal ohmic-inductive self impedances $\underline{z}_{i,i}$ and mutual impedances $\underline{z}_{i,j}$ of n conductors (e.g. six as in the present case) can be computed, considering also the electromagnetic field inside the earth; once $\underline{z}_{i,i}$ and $\underline{z}_{i,j}$ have been computed, it is possible to form the matrix $\underline{\boldsymbol{Z}}_L$ (6 × 6) and to characterize, by means of (5.1), the steady state regime of longitudinal block L of Figure 5.3 (where the voltage column vectors $\underline{\boldsymbol{u}}_S$, $\underline{\boldsymbol{u}}_R$ and the current column vectors $\underline{\boldsymbol{i}}_S$, $\underline{\boldsymbol{i}}_{SL}$, $\underline{\boldsymbol{i}}_{ST}$,

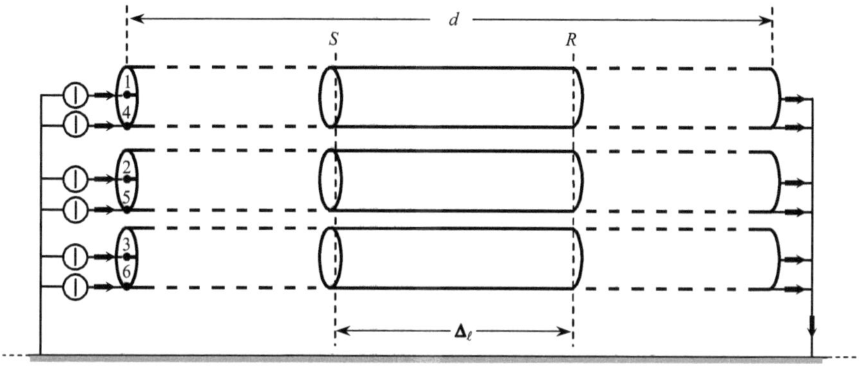

Figure 5.2 Three single-core cables with earth return

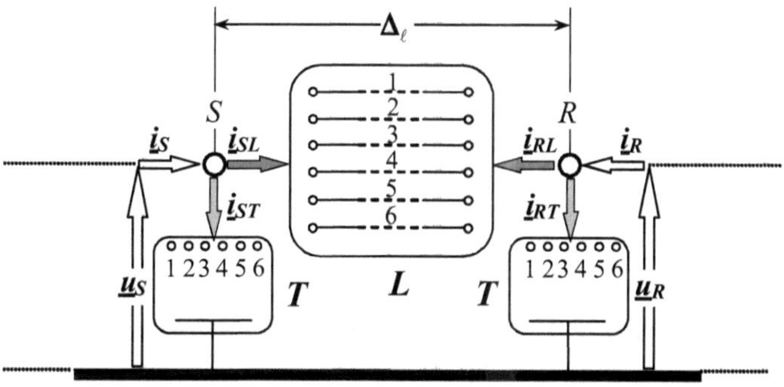

Figure 5.3 Elementary cell (of length Δ_ℓ) for the three single-core cables

$\underline{i}_R, \underline{i}_{RL}, \underline{i}_{RT}$ are shown):

$$\underline{u}_S - \underline{u}_R = \underline{Z}_L \underline{i}_{SL} \ , \tag{5.1}$$

and by considering the evident relation (5.2)

$$\underline{i}_{RL} \equiv -\underline{i}_{SL} \ , \tag{5.2}$$

it yields, (being $\underline{Z}_L$ not singular)

$$\underline{Z}_L^{-1} \underline{u}_S - \underline{Z}_L^{-1} \underline{u}_R = \underline{i}_{SL} \ , \tag{5.3}$$

$$-\underline{Z}_L^{-1} \underline{u}_S + \underline{Z}_L^{-1} \underline{u}_R = \underline{i}_{RL} \ , \tag{5.4}$$

and hence the following matrix relation (5.5) where $\underline{Y}_{L\Delta}$ (12×12) regards the block L circuit formed by the six longitudinal links:

$$\underset{\substack{\underline{i}_{L\Delta} \\ (12\times 1)}}{\begin{bmatrix} \underline{i}_{SL} \\ \underline{i}_{RL} \end{bmatrix}} = \underset{\substack{\underline{Y}_{L\Delta} \\ (12\times 12)}}{\begin{bmatrix} \underline{Z}_L^{-1} & -\underline{Z}_L^{-1} \\ -\underline{Z}_L^{-1} & \underline{Z}_L^{-1} \end{bmatrix}} \underset{\substack{\underline{u}_{\Delta} \\ (12\times 1)}}{\begin{bmatrix} \underline{u}_S \\ \underline{u}_R \end{bmatrix}} \ . \tag{5.5}$$

In particular, it is important to mark the directions of the currents in correspondence to S and R (both towards the circuital block) since the study will be developed by means of models identified by nodal admittance matrices.

Being Δ_ℓ sufficiently small, it is possible to lump the uniformly distributed shunt admittances at both ends of the cell (transverse blocks T_S and T_R) and to consider separately the longitudinal elements in the block L.

The vectors of the shunt currents at the sending-end $\underline{i}_{ST}$ and at the receiving-end $\underline{i}_{RT}$ are

$$\underbrace{\begin{bmatrix}\underline{i}_{ST}\\ \underline{i}_{RT}\end{bmatrix}}_{\substack{\underline{i}_{T\Delta}\\ (12\times1)}} = \underbrace{\begin{bmatrix}\underline{Y}_{TS} & \\ & \underline{Y}_{TR}\end{bmatrix}}_{\substack{\underline{Y}_{T\Delta}\\ (12\times12)}} \underbrace{\begin{bmatrix}\underline{u}_{S}\\ \underline{u}_{R}\end{bmatrix}}_{\substack{\underline{u}_{\Delta}\\ (12\times1)}}, \tag{5.6}$$

where $\underline{Y}_{TS}$ (6×6) and $\underline{Y}_{TR}$ (6×6) are the admittance matrices of each transversal block T, which can be computed immediately (see Section 5.2.5) if the susceptances distributed on $\Delta_\ell/2$ are lumped at each end (see Figure 5.3).

5.2.1 The Admittance Matrix $\underline{Y}_\Delta$ to Model the Elementary Cell

The superimposition of (5.5) and (5.6) yields

$$\underbrace{\begin{bmatrix}\underline{i}_{S}\\ \underline{i}_{R}\end{bmatrix}}_{\substack{\underline{i}_{\Delta}\\ (12\times1)}} = \underbrace{\begin{bmatrix}\underline{Y}_{L\Delta}+\underline{Y}_{T\Delta}\end{bmatrix}}_{\substack{\underline{Y}_{\Delta}\\ (12\times12)}} \underbrace{\begin{bmatrix}\underline{u}_{S}\\ \underline{u}_{R}\end{bmatrix}}_{\substack{\underline{u}_{\Delta}\\ (12\times1)}} \tag{5.7}$$

which completely represents the steady state regime of the elementary cell (of length Δ_ℓ). It is important to point out that in the above (5.7) the suitable matrix and vector partitions have been introduced: they allow highlighting and methodically grouping together the steady state regimes in S and in R; it will be shown (once chosen Δ_ℓ sufficiently small, not greater than 0.3 km) how this model of the elementary cell with its admittance matrix $\underline{Y}_\Delta$ can be repeatedly used together with the cross-bonding modelling to realize cascade connections and to simulate multiconductor systems of great length.

Figure 5.4a,b Possible configurations of i and j conductors

5.2.2 Computation of $\underline{Z}_L$ by Means of Simplified Carson–Clem Formulae

If $d_{ij} \leq 0.135 D_{\mathrm{CA}}$ (see Vol. II, page 154 in [1]), the self and mutual longitudinal impedances can be obtained through the Carson–Clem's formulae (5.8) and (5.9):

$$\underline{z}_{i,i} = r_i + \pi^2 \cdot 10^{-4} \cdot f + j \cdot 4\pi \cdot 10^{-4} \cdot f \cdot \ln\left(\frac{2D_{\mathrm{CA}}}{d_i'}\right) \quad [\Omega/\mathrm{km}]\,, \tag{5.8}$$

$$\underline{z}_{i,j} = \pi^2 \cdot 10^{-4} \cdot f + j \cdot 4\pi \cdot 10^{-4} \cdot f \cdot \ln\left(\frac{D_{\mathrm{CA}}}{d_{ij}}\right) \quad [\Omega/\mathrm{km}]\,, \tag{5.9}$$

where r_i = resistance per unit length of conductor i (the skin and proximity effects can be computed by means of IEC 60287 [10]; it is worth remembering that the Milliken conductors, if enamelled, dramatically reduce them); $D_{\mathrm{CA}} = 660\sqrt{\rho_g/f}$ [m] "Carson Depth"; f = power frequency, 50–60 [Hz]; ρ_g = electrical resistivity of the soil [Ω m]; $d_i' = 2$ GMR [m] (Geometrical Mean Radius); usually in EHV XLPE-cables the conductor is hollow Milliken type; for GMR, see Vol. II, page 155 in [1]; d_{ij} = mutual distance [m] between conductors i and j (Figure 5.4a).

When the conductors i and j are coaxial ones, it yields $x_{j,j} \cong x_{i,j} \equiv x_{j,i}$ where j is the external conductor and i the inner one (see Figure 5.4b).

5.2.3 Computation of $\underline{Z}_L$ by Means of Complete Carson Formulae

If $d_{ij} > 0.135 D_{\mathrm{CA}}$, self $\underline{z}_{i,i}$ and mutual $\underline{z}_{i,j}$ impedances in $\underline{\boldsymbol{Z}}_L$ must be computed by means of the following complete Carson's formulae (which have demonstrated its validity also at high frequency for distribution line carrier analysis [11]):

$$\underline{z}_{i,i} = (r_i + \Delta R_i) + \mathrm{j}\left(\omega \frac{\mu_0}{2\pi} \ln \frac{2h_i}{\mathrm{GMR}} + \Delta X_i\right) \quad [\Omega/\mathrm{km}]\,, \tag{5.10}$$

$$\underline{z}_{i,j} = \underline{z}_{j,i} = \Delta R_{ij} + \mathrm{j}\left(\omega \frac{\mu_0}{2\pi} \ln \frac{2D_{ij}}{d_{ij}} + \Delta X_{ij}\right) \quad [\Omega/\mathrm{km}]\,, \tag{5.11}$$

where Carson's correction terms ΔR_i, ΔX_i and ΔR_{ij}, ΔX_{ij} are computed by means of series [12]; h_i is the height above ground-level of conductor i and has a negative value for an underground conductor. D_{ij} is the distance between conductor i and the image of conductor j or *vice versa*.

Figure 5.5 "Impedance equivalent circuit" [9]

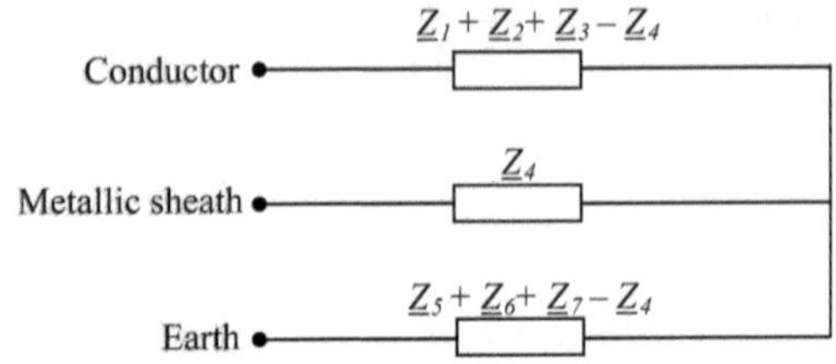

5.2.4 Computation of $\underline{\mathbf{Z}}_L$ After Wedepohl

Another mean to compute $\underline{\mathbf{Z}}_L$ is the Wedepohl's theory which gives approximated formulae both of Schelkunhoff loop theory [13] and of Pollaczek one [7] and which allows forming the matrix $\underline{\mathbf{Z}}_{LW}$ differing from $\underline{\mathbf{Z}}_L$ only for a different order of the phases and sheaths as in (5.12):

$\underline{Z}_{LW}=$ 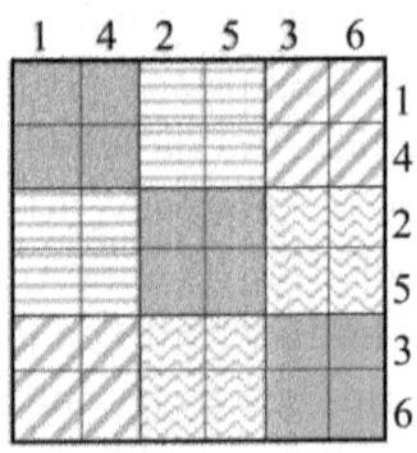 (5.12)

In order to calculate each element of the submatrices, the formulae are detailed in [9] so that it is immediate to obtain $\underline{\mathbf{Z}}_L$ from $\underline{\mathbf{Z}}_{LW}$. In particular, each diagonal submatrix (full grey in (5.12)) is constructed from the seven impedances $\underline{Z}_1 \ldots \underline{Z}_7$ and can be derived from the "impedance equivalent circuit" of Figure 5.5.

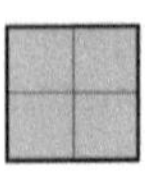

$$= \begin{bmatrix} \underline{Z}_1 + \underline{Z}_2 + \underline{Z}_3 + \underline{Z}_5 + \underline{Z}_6 + \underline{Z}_7 - 2\underline{Z}_4 & \underline{Z}_5 + \underline{Z}_6 + \underline{Z}_7 - \underline{Z}_4 \\ \hline \underline{Z}_5 + \underline{Z}_6 + \underline{Z}_7 - \underline{Z}_4 & \underline{Z}_5 + \underline{Z}_6 + \underline{Z}_7 \end{bmatrix} \begin{matrix} \textit{Ph.} \\ \textit{Sh.} \end{matrix} .$$

(columns: *Phase*, *Sheath*)

The off-diagonal submatrices in (5.12) take account of the mutual inductances between cables and may be calculated by suitably approximating the Pollaczek's formulae [7].

5.2.5 Computation of $\underline{\mathbf{Y}}_{T\Delta}$

The two matrices $\underline{\mathbf{Y}}_{TS} \equiv \underline{\mathbf{Y}}_{TR}$ can be computed taking into account the shunt capacitive and conductive links.

The kilometric admittances which must be evaluated (see Figure 5.6) are expressed by the following two formulae (see Sections 2.3.3 and 2.3.4)

$$\underline{y}_{\mathrm{a}} = g_{\mathrm{a}} + \mathrm{j}\,\omega c_{\mathrm{a}} = 10^3\,[\omega \cdot 2\pi \cdot \varepsilon_{\mathrm{a}} \cdot (\tan\delta_{\mathrm{a}} + \mathrm{j})]/\ln\left(\frac{d_{0a}}{d_{1a}}\right) \quad (\mathrm{S/km})\,, \qquad (5.13)$$

$$\underline{y}_{\mathrm{b}} = g_{\mathrm{b}} + \mathrm{j}\,\omega c_{\mathrm{b}} = 10^3\,[\omega \cdot 2\pi \cdot \varepsilon_{\mathrm{b}} \cdot (\tan\delta_{\mathrm{b}} + \mathrm{j})]/\ln\left(\frac{d_{0b}}{d_{1b}}\right) \quad (\mathrm{S/km})\,, \qquad (5.14)$$

in which g_{a} and g_{b} represent the leakage conductances across inner and outer insulations, ε_{a} and ε_{b} [F/m] the corresponding permittivities.

For instance, the admittance matrices $\underline{\boldsymbol{Y}}_{TS} \equiv \underline{\boldsymbol{Y}}_{TR}$ of the T blocks (which are easily deduced by the inspection method) are given by:

$$\underline{\boldsymbol{Y}}_{TS} = \underline{\boldsymbol{Y}}_{TR} = \begin{array}{c|c|c|c|c|c|c|}
 & 1 & 2 & 3 & 4 & 5 & 6 \\ \hline
1 & \underline{y}_{\mathrm{a}} & & & -\underline{y}_{\mathrm{a}} & & \\ \hline
2 & & \underline{y}_{\mathrm{a}} & & & -\underline{y}_{\mathrm{a}} & \\ \hline
3 & & & \underline{y}_{\mathrm{a}} & & & -\underline{y}_{\mathrm{a}} \\ \hline
4 & -\underline{y}_{\mathrm{a}} & & & \underline{y}_{\mathrm{a}} + \underline{y}_{\mathrm{b}} & & \\ \hline
5 & & -\underline{y}_{\mathrm{a}} & & & \underline{y}_{\mathrm{a}} + \underline{y}_{\mathrm{b}} & \\ \hline
6 & & & -\underline{y}_{\mathrm{a}} & & & \underline{y}_{\mathrm{a}} + \underline{y}_{\mathrm{b}} \\ \hline
\end{array} \cdot \frac{\Delta\ell}{2}\,. \qquad (5.15)$$

When the installation is not completely embedded in soil, a precise computation of the self and mutual admittances regarding the sheaths would be problematic but not so important since the sheaths are subjected to very low voltages on average along the route due to the earthing and the short-circuit links of the cross-bonding: on the basis of this consideration it appears licit to neglect $\underline{y}_{\mathrm{b}}$ in (5.15).

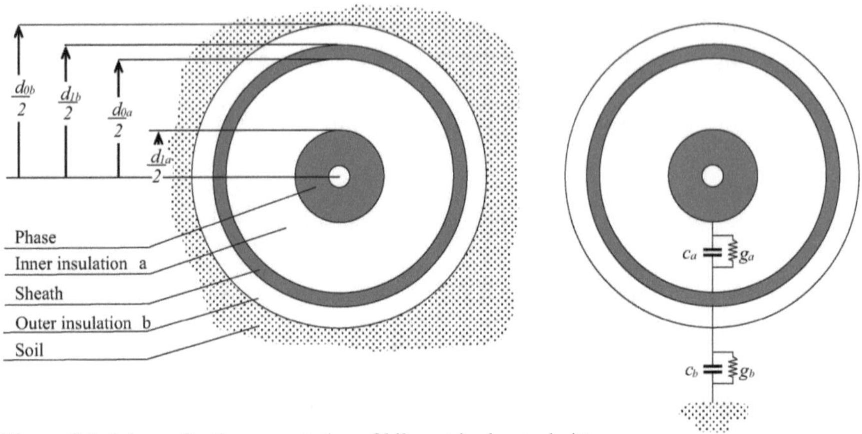

Figure 5.6 Scheme for the computation of kilometric shunt admittances

5.3 Transposition Joints Modelling: $\underline{Y}_J$

The basic idea for representing the transpositions given by the joints at minor section ends is to form their corresponding admittance matrices $\underline{Y}_J$ having purely conductive elements.

Figure 5.7 shows the links which perform the transpositions of phases and sheaths due to the joints in a, b, c of Figure 5.1 and their corresponding admittance "joint matrix" $\underline{Y}_J$.

In order to make the model more real as possible, the conductance G imputed to the links between S_x and S_y should assume values in accordance with the constructive characteristics of the joints.

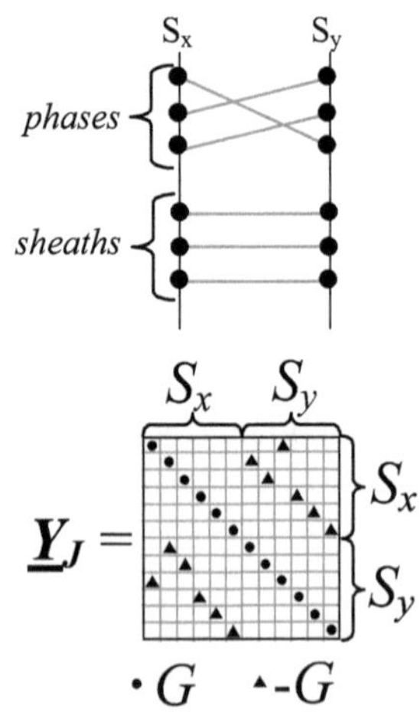

Figure 5.7 Admittance matrix $\underline{Y}_J$ for cross-bonding joints

Indicatively (manufactures' data lacking) it can be assumed (for phase and sheath links) a conductance G of about 5 kS.

5.4 Earthing of Sheaths and Insertion of Possible Shunt Reactors: $\boldsymbol{Y}_{\mathrm{E}}$; $\underline{\boldsymbol{Y}}_{\mathrm{E}\xi}$

In order to complete the model of the cross-bonded circuit it is necessary to form the matrix $\boldsymbol{Y}_{\mathrm{E}}$ which represents the bonding of the sheaths (between themselves by means of a resistance r_{E} and with the earth by means of a cross-bonded box resistance R_{E}) at the ends of each major section (see Figure 5.1).

There are two procedures in order to obtain the admittance matrix $\boldsymbol{Y}_{\mathrm{E}}$.

The first direct procedure is based on the inspection method which gives the following matrix:

$\boldsymbol{Y}_{\mathrm{E}} =$

4	5	6	
$\frac{r_{\mathrm{E}}+2R_{\mathrm{E}}}{r_{\mathrm{E}}(r_{\mathrm{E}}+3R_{\mathrm{E}})}$	$\frac{-R_{\mathrm{E}}}{r_{\mathrm{E}}(r_{\mathrm{E}}+3R_{\mathrm{E}})}$	$\frac{-R_{\mathrm{E}}}{r_{\mathrm{E}}(r_{\mathrm{E}}+3R_{\mathrm{E}})}$	4
$\frac{-R_{\mathrm{E}}}{r_{\mathrm{E}}(r_{\mathrm{E}}+3R_{\mathrm{E}})}$	$\frac{r_{\mathrm{E}}+2R_{\mathrm{E}}}{r_{\mathrm{E}}(r_{\mathrm{E}}+3R_{\mathrm{E}})}$	$\frac{-R_{\mathrm{E}}}{r_{\mathrm{E}}(r_{\mathrm{E}}+3R_{\mathrm{E}})}$	5
$\frac{-R_{\mathrm{E}}}{r_{\mathrm{E}}(r_{\mathrm{E}}+3R_{\mathrm{E}})}$	$\frac{-R_{\mathrm{E}}}{r_{\mathrm{E}}(r_{\mathrm{E}}+3R_{\mathrm{E}})}$	$\frac{r_{\mathrm{E}}+2R_{\mathrm{E}}}{r_{\mathrm{E}}(r_{\mathrm{E}}+3R_{\mathrm{E}})}$	6

The second procedure requires to form first the impedance matrix $\boldsymbol{Z}_{\mathrm{E}}$ (3×3) as seen at the nodes 4, 5 and 6 (corresponding to the sheaths), in which the self and mutual resistances can be immediately expressed as a function of r_{E} (resistance of few mΩ of the bonding links) and of R_{E} (earth resistance):

$\boldsymbol{Z}_{\mathrm{E}} =$

4	5	6	
$R_{\mathrm{E}} + r_{\mathrm{E}}$	R_{E}	R_{E}	4
R_{E}	$R_{\mathrm{E}} + r_{\mathrm{E}}$	R_{E}	5
R_{E}	R_{E}	$R_{\mathrm{E}} + r_{\mathrm{E}}$	6

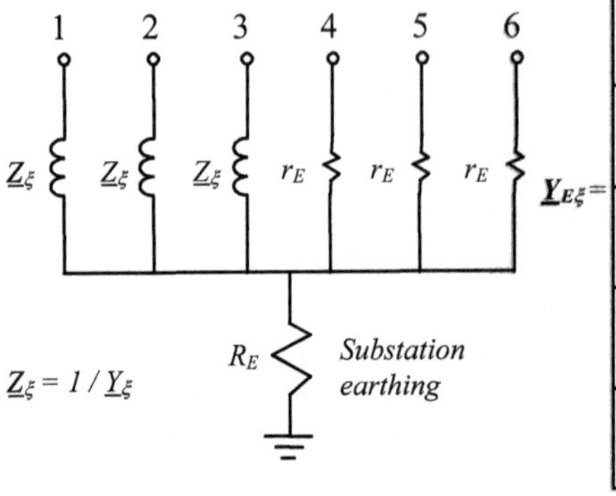

$$\underline{\boldsymbol{Y}}_{E\xi}=\begin{bmatrix} \underline{y}_{11} & \underline{y}_{12} & \underline{y}_{13} & \underline{y}_{14} & \underline{y}_{15} & \underline{y}_{16} \\ \underline{y}_{21} & \underline{y}_{22} & \underline{y}_{23} & \underline{y}_{24} & \underline{y}_{25} & \underline{y}_{26} \\ \underline{y}_{31} & \underline{y}_{32} & \underline{y}_{33} & \underline{y}_{34} & \underline{y}_{35} & \underline{y}_{36} \\ \underline{y}_{41} & \underline{y}_{42} & \underline{y}_{43} & \underline{y}_{44} & \underline{y}_{45} & \underline{y}_{46} \\ \underline{y}_{51} & \underline{y}_{52} & \underline{y}_{53} & \underline{y}_{54} & \underline{y}_{55} & \underline{y}_{56} \\ \underline{y}_{61} & \underline{y}_{62} & \underline{y}_{63} & \underline{y}_{64} & \underline{y}_{65} & \underline{y}_{66} \end{bmatrix}$$

$$\underline{y}_{11}=\underline{y}_{22}=\underline{y}_{33}=\underline{Y}_{\xi}\frac{r_E+3R_E+2r_ER_E\underline{Y}_{\xi}}{r_E+3R_E+3r_ER_E\underline{Y}_{\xi}}$$

$$\underline{y}_{44}=\underline{y}_{55}=\underline{y}_{66}=\frac{1}{r_E}\cdot\frac{r_E+2R_E+3r_ER_E\underline{Y}_{\xi}}{r_E+3R_E+3r_ER_E\underline{Y}_{\xi}}$$

$$\underline{y}_{12}=\underline{y}_{13}=\underline{y}_{21}=\underline{y}_{31}=\underline{y}_{32}=\underline{y}_{23}=-\underline{Y}_{\xi}^{2}\cdot\frac{r_ER_E}{r_E+3R_E+3r_ER_E\underline{Y}_{\xi}}$$

$$\underline{y}_{45}=\underline{y}_{54}=\underline{y}_{64}=\underline{y}_{46}=\underline{y}_{65}=\underline{y}_{56}=\frac{1}{r_E}\cdot\frac{r_E+2R_E+3r_ER_E\underline{Y}_{\xi}}{r_E+3R_E+3r_ER_E\underline{Y}_{\xi}}$$

$$\left.\begin{array}{l}\underline{y}_{14}=\underline{y}_{15}=\underline{y}_{16}=\underline{y}_{41}=\underline{y}_{51}=\underline{y}_{61}\\ \underline{y}_{24}=\underline{y}_{25}=\underline{y}_{26}=\underline{y}_{42}=\underline{y}_{52}=\underline{y}_{62}\\ \underline{y}_{34}=\underline{y}_{35}=\underline{y}_{36}=\underline{y}_{43}=\underline{y}_{53}=\underline{y}_{63}\end{array}\right\}=-\underline{Y}_{\xi}\cdot\frac{R_E}{r_E+3R_E+3r_ER_E\underline{Y}_{\xi}}$$

Figure 5.8 Admittance matrix $\underline{\boldsymbol{Y}}_{E\xi}$ for shunt reactors and sheath earthing

The inversion of $\boldsymbol{Z}_E$ gives immediately the matrix $\boldsymbol{Y}_E$ (a purely conductive one)

$$\boldsymbol{Y}_E=\boldsymbol{Z}_E^{-1}$$

which will be introduced (as it will be shown) in each major section end in the modelling of the whole system.

It is worth noting that the possible presence of lumped shunt reactive compensation (insertion between the phases 1, 2, 3 and the earth) by means of three single-phase reactors can be easily accounted for by means of the admittance matrix $\underline{\boldsymbol{Y}}_{E\xi}$ to be overlapped (as detailed in the example of Section 5.9) in the right locations.

Analogously to $\boldsymbol{Y}_E$, there are two methods in order to compute $\underline{\boldsymbol{Y}}_{E\xi}$: a direct calculation by the inspection method as in Figure 5.8.

There is also the possibility of computing firstly the matrix $\underline{\boldsymbol{Z}}_{\mathrm{E}\xi}$ (6×6), where $\underline{Z}_{\xi} = 1/\underline{Y}_{\xi}$,

$$\underline{\boldsymbol{Z}}_{\mathrm{E}\xi} = \begin{bmatrix} \underline{Z}_{\xi} + R_{\mathrm{E}} & R_{\mathrm{E}} & R_{\mathrm{E}} & R_{\mathrm{E}} & R_{\mathrm{E}} & R_{\mathrm{E}} \\ R_{\mathrm{E}} & \underline{Z}_{\xi} + R_{\mathrm{E}} & R_{\mathrm{E}} & R_{\mathrm{E}} & R_{\mathrm{E}} & R_{\mathrm{E}} \\ R_{\mathrm{E}} & R_{\mathrm{E}} & \underline{Z}_{\xi} + R_{\mathrm{E}} & R_{\mathrm{E}} & R_{\mathrm{E}} & R_{\mathrm{E}} \\ R_{\mathrm{E}} & R_{\mathrm{E}} & R_{\mathrm{E}} & r_{\mathrm{E}} + R_{\mathrm{E}} & R_{\mathrm{E}} & R_{\mathrm{E}} \\ R_{\mathrm{E}} & R_{\mathrm{E}} & R_{\mathrm{E}} & R_{\mathrm{E}} & r_{\mathrm{E}} + R_{\mathrm{E}} & R_{\mathrm{E}} \\ R_{\mathrm{E}} & R_{\mathrm{E}} & R_{\mathrm{E}} & R_{\mathrm{E}} & R_{\mathrm{E}} & r_{\mathrm{E}} + R_{\mathrm{E}} \end{bmatrix}$$

and then the admittance matrix $\underline{\boldsymbol{Y}}_{\mathrm{E}\xi} = (\underline{\boldsymbol{Z}}_{\mathrm{E}\xi})^{-1}$.

5.5 The Multiconductor Supply Model at the Sending-End

It is worth mentioning that in Chapter 3 (where positive sequence models have always been used) the steady regime had been studied by applying a voltage supply of $U_{0S} = 230\,\mathrm{kV}$ directly at port S, whereas for the analysis of the no-load energization (after Section 3.7) an equivalent subtransient impedance $\mathrm{j}\,X''$ of the supply network had been considered: analogously, for the study of multiconductor steady state regimes, the following positive sequence voltages must be directly applied

$$\underline{\boldsymbol{u}}_{\mathrm{ph}} = \begin{bmatrix} 1 \\ \alpha^2 \\ \alpha \end{bmatrix} \cdot 230\,\mathrm{kV}\,, \quad \text{where } \alpha = \mathrm{e}^{\mathrm{j}\,\frac{2}{3}\pi}\,,$$

and for the study of the multiconductor no-load energization, also a phase model of the supply network must be introduced by means of the matrix $\underline{\boldsymbol{Z}}_{\mathrm{ph}}$ (3×3) of the self and mutual longitudinal phase impedances.

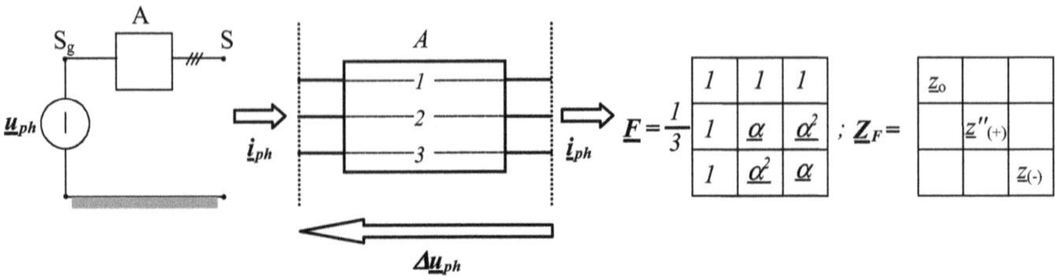

Figure 5.9 Supply modelling (in subtransient regimes)

Since in the subtransient regimes the equivalent three-phase supply network (block A of Figure 5.9) is usually hypothesized as a symmetric structure, it can be fully characterized by the corresponding zero, (subtransient) positive and negative longitudinal impedances (after Fortescue [14]) whose values are coherently computed from the TSO in the power network studies and give rise to the sequence matrix $\underline{Z}_{\mathrm{F}}$.

The corresponding phase matrix $\underline{Z}_{\mathrm{ph}}$ (full matrix composed of the self and mutual longitudinal impedances of and between the phases) can be obtained (see analogous examples in [15]) by means of (5.16)–(5.18) which make use of the Fortescue transformation $\underline{F}$ (with $\Delta\underline{u}_{\mathrm{F}}$ and $\underline{i}_{\mathrm{F}}$ consisting of sequence phasors)

$$\Delta\underline{u}_{\mathrm{F}} = \underline{F}\,\Delta\underline{u}_{\mathrm{ph}}\;; \qquad \underline{i}_{\mathrm{F}} = \underline{F}\,\underline{i}_{\mathrm{ph}}\;; \qquad \Delta\underline{u}_{\mathrm{F}} = \underline{Z}_{\mathrm{F}}\underline{i}_{\mathrm{F}}\;; \tag{5.16}$$

$$\underline{F}\,\Delta\underline{u}_{\mathrm{ph}} = \underline{Z}_{\mathrm{F}}\underline{i}_{\mathrm{F}} \;\rightarrow\; \Delta\underline{u}_{\mathrm{ph}} = \underline{F}^{-1}\underline{Z}_{\mathrm{F}}\underline{i}_{\mathrm{F}} \;\rightarrow\; \Delta\underline{u}_{\mathrm{ph}} = \underline{F}^{-1}\underline{Z}_{\mathrm{F}}\underline{F}\,\underline{i}_{\mathrm{ph}}\,, \tag{5.17}$$

whence

$$\underline{Z}_{\mathrm{ph}} = \underline{F}^{-1}\underline{Z}_{\mathrm{F}}\underline{F}\,. \tag{5.18}$$

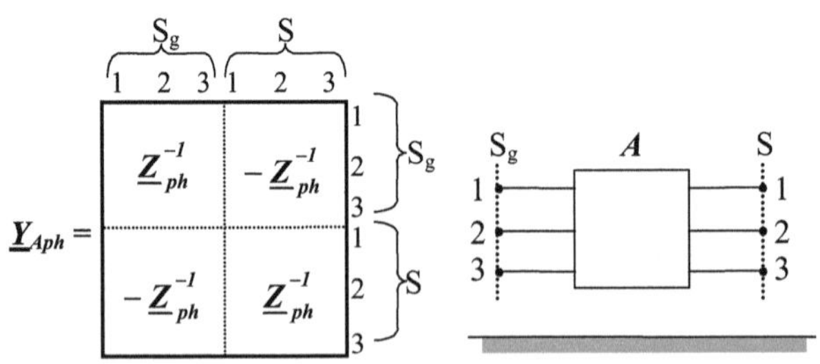

Figure 5.10 Subtransient admittance matrix of supply model (multiconductor)

By following the same criteria used for the longitudinal block L of Section 5.2 it is possible (see Figure 5.10) to achieve for the block A the matrix $\underline{Y}_{A\text{ph}}$ which characterizes the power supply in the sub-transient regimes.

5.6 Equivalent Receiving-End Matrix for Load Modelling

In order to model a complex power loading at receiving-end, a three-phase balanced static load can be considered as in Figure 5.11.

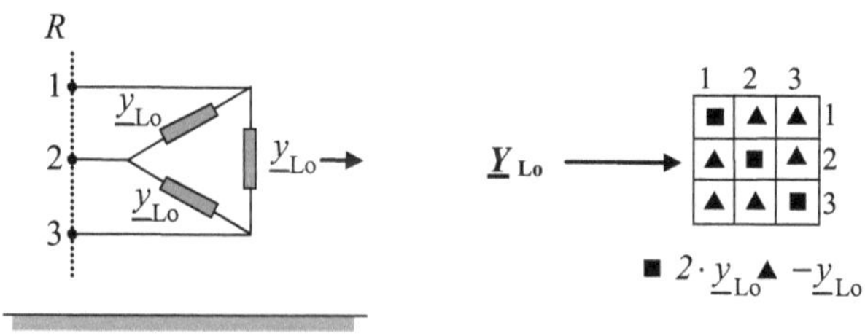

Figure 5.11 Model of complex power demand at receiving-end

It is worth noting that the UGC asymmetry cannot be totally zeroed, even if drastically reduced from cross-bonding with phase transpositions: therefore even with a perfectly symmetric power supply it has some voltage asymmetry at R.

Notwithstanding, the value of $\underline{y}_{\text{Lo}}$ can be computed after (5.19) with symmetrical phase to phase voltages with magnitude U (e.g. $U = \sqrt{3}U_{0R}$ already computed with Ossanna's method for the same hypothesized complex power $\underline{S}_{\text{Lo}}$ in the positive sequence modelling).

$$\underline{y}_{\text{Lo}} = \frac{\underline{S}_{\text{Lo}}}{3 \cdot U^2}. \tag{5.19}$$

5.7 The Cascade Composition of Blocks Modelled by "Admittance Partitioned Matrices": A First, Simple Circuit

In order to give a clear exposition of how to compose computationally different network elements modelled by their partitioned admittance matrices let us consider the example of Figure 5.12, where the blocks a, b, c are some elementary cells of UGC modelled by the matrices, each with dimensions 12×12, $\underline{Y}_a$, $\underline{Y}_b$, $\underline{Y}_c$ given by (5.7) in Section 5.2.1. On the basis of (5.20)–(5.22) which lead to the general formulation of the current vectors injected in the different sections 1, 2, 3, 4, it can be ascertained that the proposed system is expressible by (5.23), where the matrix $\underline{Y}$ is formed of the partial overlapping ("tile style") of the matrices $\underline{Y}_a$, $\underline{Y}_b$, $\underline{Y}_c$.

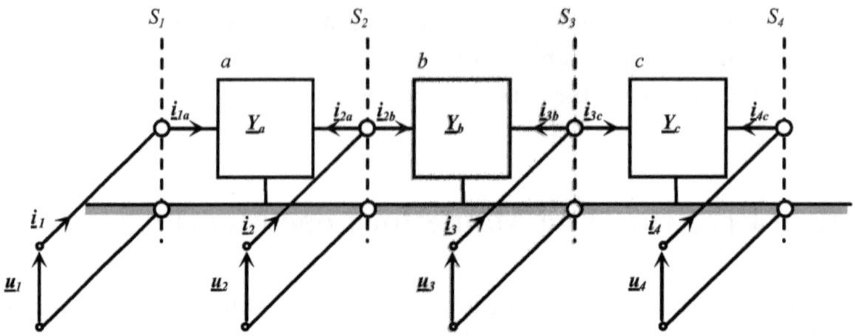

Figure 5.12 Cascade composition of multiconductor cells

$$\underline{i}_1 = \underline{i}_{1a}, \quad \underline{i}_2 = \underline{i}_{2a} + \underline{i}_{2b}, \quad \underline{i}_3 = \underline{i}_{3b} + \underline{i}_{3c}, \quad \underline{i}_4 = \underline{i}_{4c}$$

$$\begin{bmatrix} \underline{i}_{1a} \\ \underline{i}_{2a} \end{bmatrix} = \underbrace{\begin{bmatrix} \underline{Y}_{a11} & \underline{Y}_{a12} \\ \underline{Y}_{a21} & \underline{Y}_{a22} \end{bmatrix}}_{\underline{Y}_a} \begin{bmatrix} \underline{u}_1 \\ \underline{u}_2 \end{bmatrix}, \tag{5.20}$$

$$\begin{bmatrix} \underline{i}_{2b} \\ \underline{i}_{3b} \end{bmatrix} = \underbrace{\begin{bmatrix} \underline{Y}_{b22} & \underline{Y}_{b23} \\ \underline{Y}_{b32} & \underline{Y}_{b33} \end{bmatrix}}_{\underline{Y}_b} \begin{bmatrix} \underline{u}_2 \\ \underline{u}_3 \end{bmatrix}, \tag{5.21}$$

$$\begin{bmatrix} \underline{i}_{3c} \\ \underline{i}_{4c} \end{bmatrix} = \underbrace{\begin{bmatrix} \underline{Y}_{c33} & \underline{Y}_{c34} \\ \underline{Y}_{c43} & \underline{Y}_{c44} \end{bmatrix}}_{\underline{Y}_c} \begin{bmatrix} \underline{u}_3 \\ \underline{u}_4 \end{bmatrix}, \tag{5.22}$$

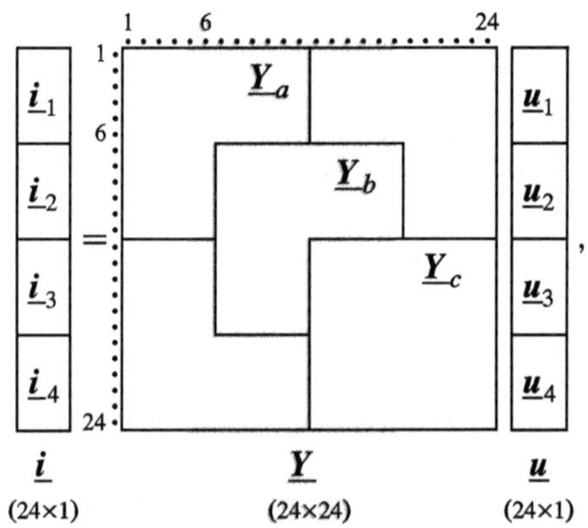

(5.23)

The classical cascade connection of blocks a, b, c becomes defined by the constrain $\underline{i}_2 = \underline{i}_3 \equiv \mathbf{0}$.

5.7.1 *The Introduction of Other Blocks in the First Simple Circuit and the Steady State Analysis*

The above mentioned simple circuit can now become more realistic if the following additions are performed (see Figure 5.13a):

- the bonding and earthing of the sheaths in section 1 (by means of the model represented by the matrix $\underline{\boldsymbol{Y}}_{\mathrm{E}}$ as in Section 5.4) realizing a single-point bonding since the cable, in this first simple example with three elementary cells, is short;
- the introduction in the receiving end of a balanced three-phase load modelled by the $\underline{\boldsymbol{Y}}_{\mathrm{Lo}}$ as in Section 5.6;
- the imposition (after Section 5.5) of the subvector $\underline{\boldsymbol{u}}_{\mathrm{ph}}$ (3×1) at the terminals 1, 2, 3 of section S_1 which causes the injection of the current subvector $\boldsymbol{i}_{\mathrm{ph}}$ (3×1).

Therefore it is possible to obtain the new relation (5.24) where:

- the matrix $\underline{\boldsymbol{Y}}$ has been updated by means of the overlapping, in the suitable positions, of the two matrices $\underline{\boldsymbol{Y}}_{\mathrm{E}}$ and $\underline{\boldsymbol{Y}}_{\mathrm{Lo}}$;

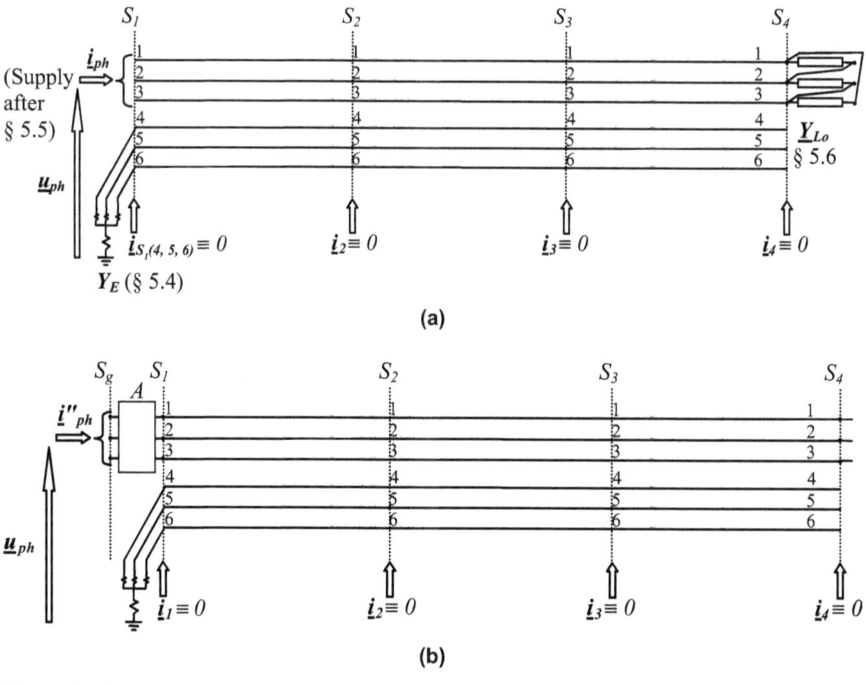

Figure 5.13 (**a**) Completion of the model in Figure 5.12. (**b**) Updated (**a**) for no-load subtransient energization

- in the structure of vector $\underline{\boldsymbol{i}}$, the subvectors $\underline{\boldsymbol{i}}_{\mathrm{ph}} \neq 0$ and $\underline{\boldsymbol{i}}_{x} \equiv \mathbf{0}$ (as highlighted by the Figure 5.13a).

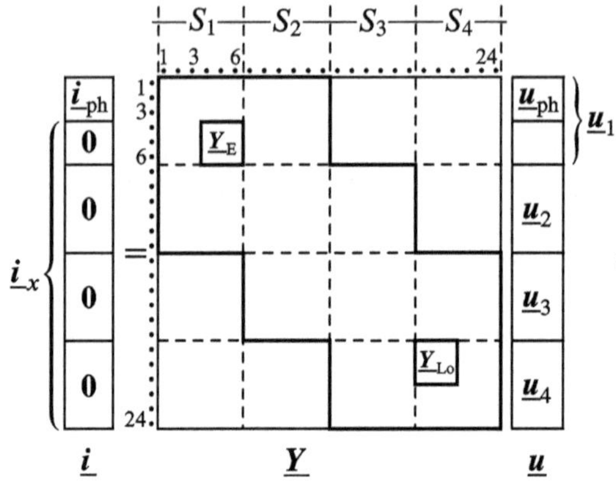

, (5.24)

Relation (5.24) simulates the supply of a load ($\underline{\boldsymbol{Y}}_{\mathrm{Lo}}$) by means of the cascade of three cells.

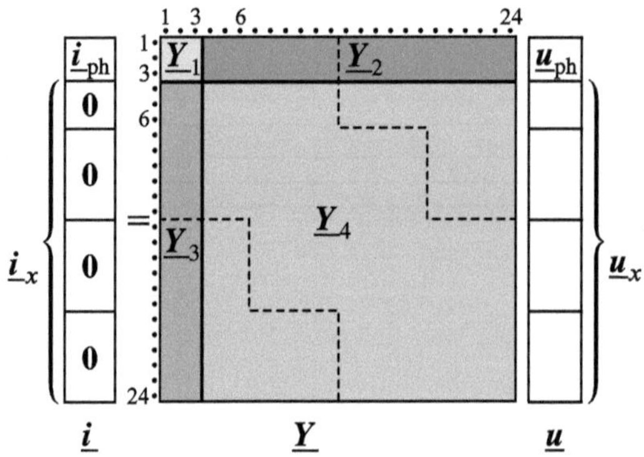

, (5.25)

By introducing suitable partitions in (5.24), as shown in (5.25), (5.26) can be obtained

$$\underline{\boldsymbol{i}}_{x} \equiv \mathbf{0} = \underline{\boldsymbol{Y}}_{3}\underline{\boldsymbol{u}}_{\mathrm{ph}} + \underline{\boldsymbol{Y}}_{4}\underline{\boldsymbol{u}}_{x} \ , \tag{5.26}$$

and thus

$$\underline{\boldsymbol{u}}_{x} = -\underline{\boldsymbol{Y}}_{4}^{-1}\underline{\boldsymbol{Y}}_{3}\underline{\boldsymbol{u}}_{\mathrm{ph}} \ , \tag{5.27}$$

where $\underline{\boldsymbol{Y}}_{4}$ is not singular and $\underline{\boldsymbol{u}}_{\mathrm{ph}}$ is fixed.

So knowing the subvector $\underline{\boldsymbol{u}}_{x}$, the steady state regime of any cell, and hence of the whole multiconductor system, is completely available.

5.7.2 The No-Load Subtransient Energization Analysis

As already mentioned in Section 5.5, in order to study the subtransient regime of energization at no-load, the following modifications must be performed on the circuit of Figure 5.13a (see Figure 5.13b):

- zeroing the matrix $\underline{\boldsymbol{Y}}_{\mathrm{Lo}}$ in order to model a null load at R;
- introducing the matrix $\underline{\boldsymbol{Y}}_{A\mathrm{ph}}$ (regarding the supply block A) between the section S_{g} and the terminals 1, 2, 3 of the section S_1;
- imposing the subvector $\underline{\boldsymbol{u}}_{\mathrm{ph}}$ at the terminals 1, 2, 3 of the section S_{g}.

Therefore, it can be derived the formulation of (5.28) where the matrix $\underline{\boldsymbol{Y}}''$ (which includes $\underline{\boldsymbol{Y}}_{A\mathrm{ph}}$) has the dimensions 27 × 27.

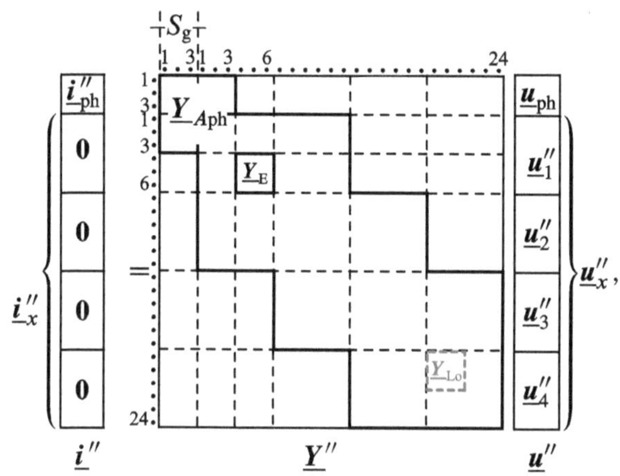

(5.28)

By introducing the suitable partition of the matrix and of the vectors, in complete analogy with (5.27), it yields

$$\underline{\boldsymbol{u}}''_x = -\underline{\boldsymbol{Y}}_4''^{-1}\underline{\boldsymbol{Y}}''_3\underline{\boldsymbol{u}}_{\mathrm{ph}}$$

and in particular the regime of the subtransient voltages at R represented by $\underline{\boldsymbol{u}}''_4$.

5.8 The Admittance Matrix Equivalent to k Blocks in Cascade Connections

When a cascade connection of multiconductor blocks is considered, it may be suitable (alternative or in addition to the procedure in Section 5.7) to use also a method which allows to synthesize a unique multiconductor block in an equivalent matrix $\underline{\boldsymbol{Y}}_{\mathrm{eq}}$, without increasing the matrix dimensions (see the matrix $\underline{\boldsymbol{Y}}$ in (5.23)).

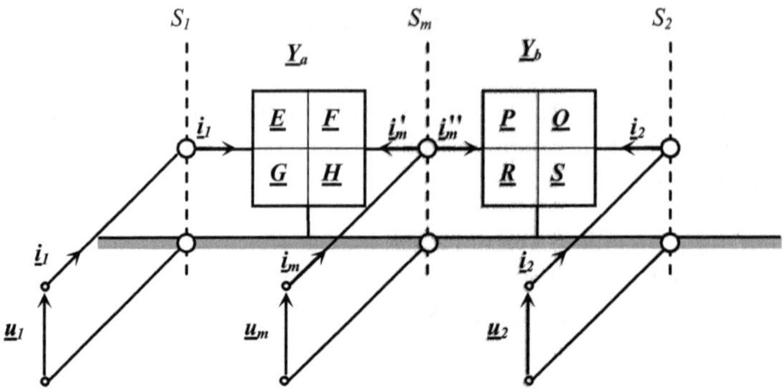

Figure 5.14 Cascade connection of two multiconductor blocks

Let us consider first the connection of two blocks of Figure 5.14 where the admittance matrices $\underline{Y}_a$ and $\underline{Y}_b$ $(2n \times 2n)$ are partitioned so that the submatrices $\underline{E} \dots \underline{H}$, $\underline{P} \dots \underline{S}$, of dimensions $n \times n$, results consistent with the dimensions of the voltage and current vectors.

By observing the current vectors injected into the system, the voltage vectors in the different sections and the partitions of the matrices $\underline{Y}_a$ and $\underline{Y}_b$, it is possible to write the following relations

$$\begin{cases} \underline{i}_1 = \underline{E}\,\underline{u}_1 + \underline{F}\,\underline{u}_\mathrm{m}\,, \\ \underline{i}'_\mathrm{m} = \underline{G}\,\underline{u}_1 + \underline{H}\,\underline{u}_\mathrm{m}\,, \end{cases} \tag{5.29}$$

$$\begin{cases} \underline{i}''_\mathrm{m} = \underline{P}\,\underline{u}_\mathrm{m} + \underline{Q}\,\underline{u}_2\,, \\ \underline{i}_2 = \underline{R}\,\underline{u}_\mathrm{m} + \underline{S}\,\underline{u}_2\,, \end{cases} \tag{5.30}$$

$$\underline{i}_\mathrm{m} = \underline{i}'_\mathrm{m} + \underline{i}''_\mathrm{m}\,, \tag{5.31}$$

whose matricial writing is also immediate:

$$\begin{bmatrix} \underline{i}_1 \\ \underline{i}_2 \\ \underline{i}_\mathrm{m} \end{bmatrix} = \begin{bmatrix} \underline{E} & \mathbf{0} & \underline{F} \\ \mathbf{0} & \underline{S} & \underline{R} \\ \underline{G} & \underline{Q} & \underline{H}+\underline{P} \end{bmatrix} \begin{bmatrix} \underline{u}_1 \\ \underline{u}_2 \\ \underline{u}_\mathrm{m} \end{bmatrix} \tag{5.32}$$

Since in the cascade connection $\underline{i}_\mathrm{m} = 0$, it has

$$\mathbf{0} = \underline{G} \cdot \underline{u}_1 + \underline{Q} \cdot \underline{u}_2 + (\underline{H} + \underline{P})\underline{u}_\mathrm{m} \tag{5.33}$$

or

$$\underline{u}_\mathrm{m} = -(\underline{H} + \underline{P})^{-1}\underline{G}\,\underline{u}_1 - (\underline{H} + \underline{P})^{-1}\underline{Q}\,\underline{u}_2 \tag{5.34}$$

which leads to the matrix relation

$$\begin{bmatrix} \underline{\boldsymbol{i}}_1 \\ \underline{\boldsymbol{i}}_2 \end{bmatrix} = \underbrace{\begin{bmatrix} \underline{\boldsymbol{E}} - \underline{\boldsymbol{F}}(\underline{\boldsymbol{H}}+\underline{\boldsymbol{P}})^{-1}\underline{\boldsymbol{G}} & -\underline{\boldsymbol{F}}(\underline{\boldsymbol{H}}+\underline{\boldsymbol{P}})^{-1}\underline{\boldsymbol{Q}} \\ -\underline{\boldsymbol{R}}(\underline{\boldsymbol{H}}+\underline{\boldsymbol{P}})^{-1}\underline{\boldsymbol{G}} & \underline{\boldsymbol{S}} - \underline{\boldsymbol{R}}(\underline{\boldsymbol{H}}+\underline{\boldsymbol{P}})^{-1}\underline{\boldsymbol{Q}} \end{bmatrix}}_{\underline{\boldsymbol{Y}}_{\mathrm{eq}}} \cdot \begin{bmatrix} \underline{\boldsymbol{u}}_1 \\ \underline{\boldsymbol{u}}_2 \end{bmatrix} \qquad (5.35)$$

$\underline{\boldsymbol{Y}}_{\mathrm{eq}}$ represents the matrix equivalent to the cascade connection of two multiconductor blocks [16].

This formulation of equivalent matrix can become very useful when the cascade of a great number k of blocks all equal and characterized by the same cell matrix $\underline{\boldsymbol{Y}}_{\Delta}$ must be considered.

For example, if each block is represented by the same $\underline{\boldsymbol{Y}}_{\Delta}$ (12×12), it is possible to compute the whole cascade matrix $\underline{\boldsymbol{Y}}_{k\Delta}$ (12×12) (as seen at the two terminal ends) with a very low number of operations even if the block number k is very high.

If for example $k = 1000$, the "cascade procedure" (applied nine times) yields at first $\underline{\boldsymbol{Y}}_{2\Delta}$; $\underline{\boldsymbol{Y}}_{4\Delta}$; $\underline{\boldsymbol{Y}}_{8\Delta}$;$\underline{\boldsymbol{Y}}_{16\Delta}$; $\underline{\boldsymbol{Y}}_{32\Delta}$; $\underline{\boldsymbol{Y}}_{64\Delta}$; $\underline{\boldsymbol{Y}}_{128\Delta}$; $\underline{\boldsymbol{Y}}_{256\Delta}$; $\underline{\boldsymbol{Y}}_{512\Delta}$.

The representation of $k = 1000$ in binary code shows that $\underline{\boldsymbol{Y}}_{1000\Delta}$ might be evaluated by applying the cascade algorithm to

$$\underline{\boldsymbol{Y}}_{512\Delta};\ \underline{\boldsymbol{Y}}_{256\Delta};\ \underline{\boldsymbol{Y}}_{128\Delta};\ \underline{\boldsymbol{Y}}_{64\Delta};\ \underline{\boldsymbol{Y}}_{32\Delta};\ \underline{\boldsymbol{Y}}_{8\Delta};$$

so that in total 14 "cascade operations" are needed [17].

5.9 Application of Multiconductor Analysis to the System "Cable #b, 60 km" Already Studied in Chapter 3 with Simplified Criteria (see Figures 3.21 and 3.36)

As already mentioned, this paragraph deals with one of the case studies of Chapter 3 i.e. cable #b (see Figure 5.15 and Table 5.1), with a length of 60 km, shunt compensated with a compensation degree $\xi_{\mathrm{sh}} = 0.608$.

With regard of the multiconductor system, Figure 5.16 (together with Figure 5.1) results extremely useful in order to visualize both the cable subdivision in elementary cells (with indication of minor and major sections) and the lumped shunt compensation locations.

In particular, Figure 5.16 shows that:

- the minor section is suitably set equal to the cable drum length of 625 m;
- each minor section is modelled as a cascade of 5 elementary cells of small length $\Delta_\ell = 125$ m, (after the recommendations of Section 5.2), each modelled by its own $\underline{\boldsymbol{Y}}_\Delta$ (after the indications of Section 5.2.1);
- at the end of each minor section, the model of cross-bonding (characterized, after Section 5.3, by the matrix $\underline{\boldsymbol{Y}}_\mathrm{J}$) must be introduced (see a, b, c of Figure 5.1);
- by observing Figure 5.1, it can be seen that each major section has a length of $3 \times 625 = 1875$ m (since it is modelled by the cascade of three minor sections included the three transposition models); at the ends of the major section the earthing of the sheaths must be taken into consideration by means of the matrices $\boldsymbol{Y}_\mathrm{E}$ suitably located, as depicted in Figure 5.1;
- the model of 60 km long UGC (without shunt compensation) gives consequently the simulation of $60/1.875 = 32$ major sections; and
- for the lumped compensation (with $\xi_{sh} = 0.608$) it is foreseen the subdivision of the line in three parts constituted by 11, 10, 11, major sections. In the suitable locations where shunt reactors are installed, the matrix $\underline{\boldsymbol{Y}}_{\mathrm{E}\xi}$ (or $\boldsymbol{Y}'_{\mathrm{E}\xi}$) must be computed: it will substitute the matrix $\boldsymbol{Y}_\mathrm{E}$ after Section 5.4.

It is worth noting that this method also suit very well the case of major sections with different lengths (as it happens in the real installations).

The model of the system must be completed with the introduction, as shown in Figure 5.13, of the balanced load modelling characterized by the matrix $\underline{\boldsymbol{Y}}_\mathrm{Lo}$ (after Section 5.6) by assuming the regime #1 of Figure 3.36 as reference.

Finally the whole matrix $\underline{\boldsymbol{Y}}$ (3456×3456) arises: after the partition of it similarly to the simple example[1] seen in Section 5.7.1, it is possible to achieve the steady state regime of UGC by means of the computation of the unknown vector $\underline{\boldsymbol{u}}_x$ due to the application of the supply vector $\underline{\boldsymbol{u}}_\mathrm{ph}$, so that the multiconductor system steady state is completely known.

Besides the regime #1 of Figure 3.36 many other regimes have been accurately investigated: they allow one to ascertain the satisfying correctness of the simplified method based on the positive sequence model.

Figure 5.17 shows, as first example, the phase current magnitudes along the cable: they are in whole accordance with those of Figure 3.37 demonstrating that the single-circuit positive sequence analysis of lumped compensation (for usual cable installations) can be sufficient for engineering purposes. Obviously, the multiconductor analysis brings to a higher accuracy of system knowledge as the zoom of Figure 5.18 (and of other curves) clearly highlights.

The zoom of Figure 5.18 shows a light exceeding of the ampacity in the first kilometres, but it is not problematic if one considers also the conservative limits due to the natural fluctuations in the line operation.

Figure 5.19 shows the phase voltage magnitudes which are almost equal to those of Figure 3.38: the scale does not allow separating the three different phases so that

[1] The inversion of the submatrix $\underline{\boldsymbol{Y}}_A$ is not problematic since it is highly sparse.

Table 5.1 Data assumed in the multiconductor computations

Multiconductor cell length		km	0.125
Cable drum (minor section)		km	0.625
Cross-bonding section length (major section)		km	1.875
Substation earthing	R_E	Ω	0.1
Earth resistivity	ρ_{soil}	Ω m	100
Cross-bonded box resistance	R_E	Ω	10
Sheath resistance at 78.2 °C (50 Hz)	r_{sh}	mΩ/km	70.0
Link resistance	r_E	mΩ	1
Shunt compensation degree	ξ_{sh}		0.608
Number of shunt compensation stations			4

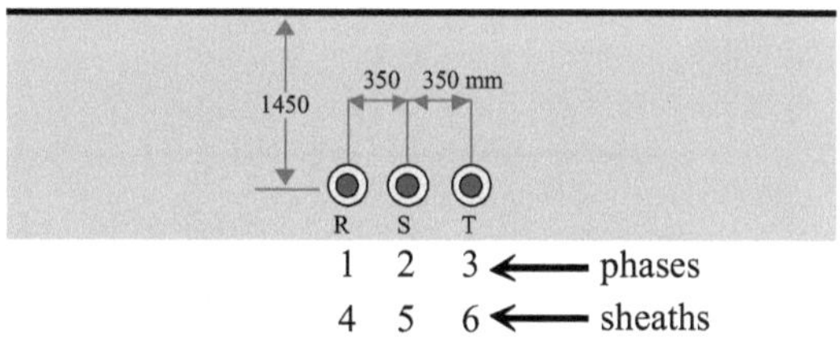

Figure 5.15 Single-circuit cable line (in cross-bonding)

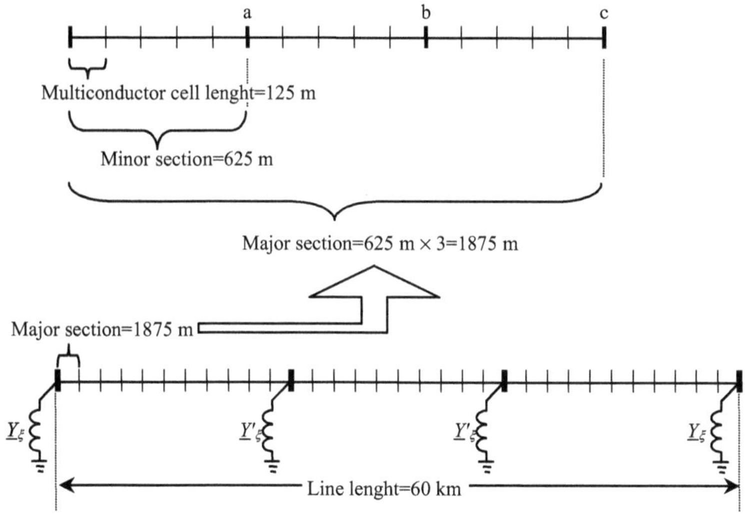

Figure 5.16 Subdivision of the single-circuit cable line in cross-bonding with indication of lumped shunt compensation locations

the zoom of Figure 5.20 shows the great accuracy and detail of the multiconductor procedure and a negligible voltage asymmetry.

One of the great possibilities offered by the multiconductor analysis is the knowledge of the sheath electric behaviour. Figure 5.21 shows the sheath voltage and current magnitudes along the major sections of the cable. The zooms in the first three major sections in Figure 5.22 show slight differences. The typical behaviour of volt-

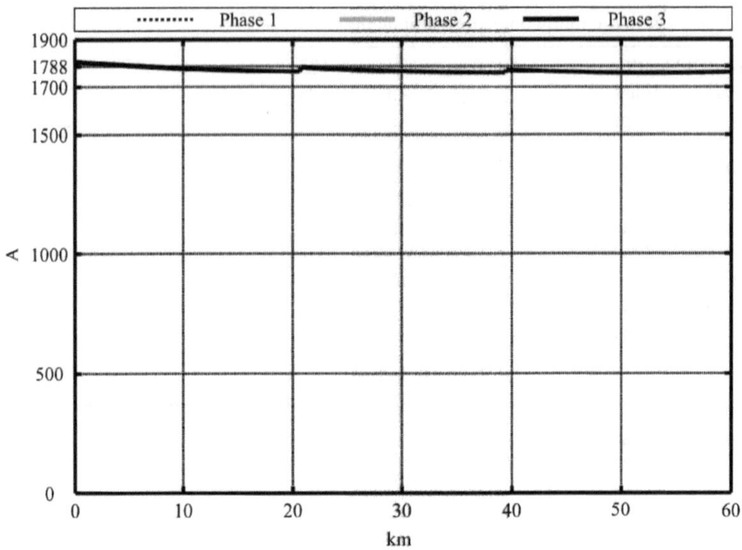

Figure 5.17 RMS phase current magnitudes along the single-circuit cable #b, ($\xi_{sh} = 0.608$ with 32 cross-bonding major sections)

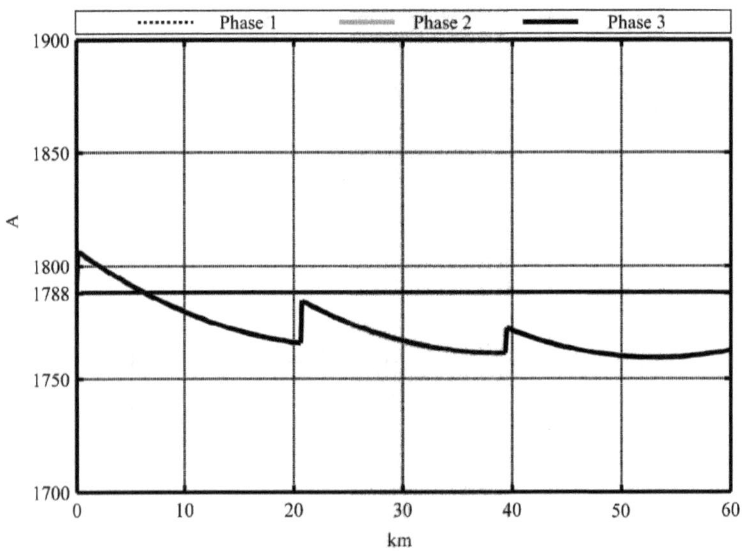

Figure 5.18 Zoom of the RMS phase current magnitudes along the single-circuit cable #b, ($\xi_{sh} = 0.608$ with 32 cross-bonding major sections)

age magnitudes can be also qualitatively and theoretically foreseen [18]. The sheath currents of induced nature are extremely low so demonstrating the effectiveness of cross-bonding arrangement: only the capacitive currents (since the multiconductor neglects the eddy currents) remain with a very complex composition.

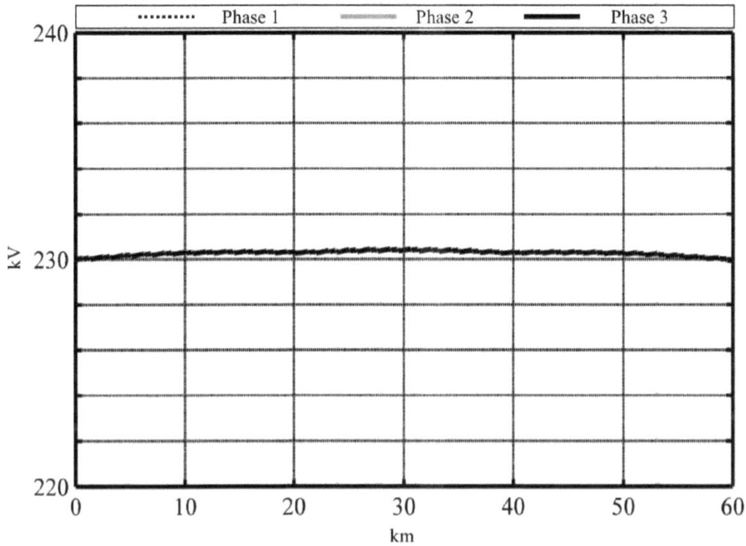

Figure 5.19 RMS phase voltage magnitudes along the single-circuit cable #b, ($\xi_{sh} = 0.608$ with 32 cross-bonding major sections)

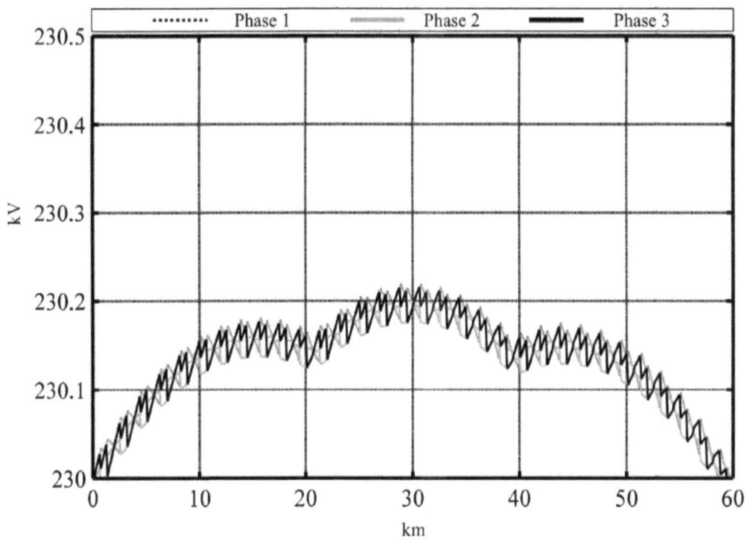

Figure 5.20 Zoom of the RMS phase voltage magnitudes along the single-circuit cable #b, ($\xi_{sh} = 0.608$ with 32 cross-bonding major sections)

As is well-known, the use of phase transposition gives a good electrical symmetrization of the flat installation so that the sheath currents result further reduced from cross-bonding with phase transpositions to that without them. The phase transposition is unnecessary in case of trefoil arrangement (which has already an intrinsic

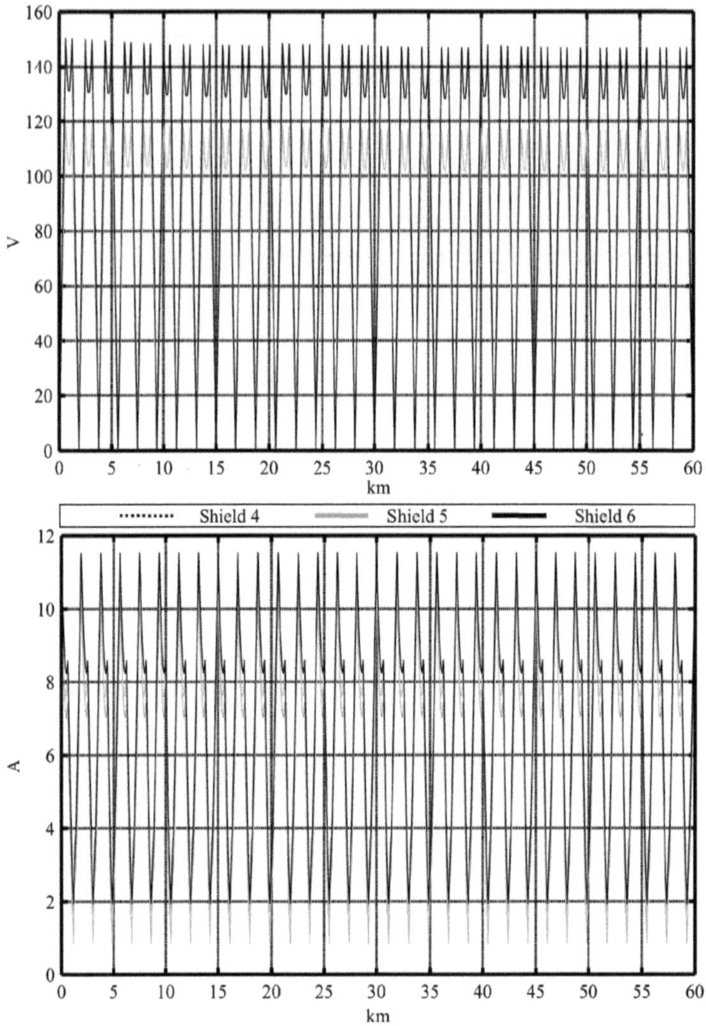

Figure 5.21 RMS sheath voltage and current magnitudes along the single-circuit cable #b, (ξ_{sh} = 0.608 with 32 cross-bonding major sections)

symmetry). An important consequence is the lower power losses in cross-bonding rather than only sheaths cross-bonded: this implies a higher ampacity.

The sheath voltage magnitudes along the line in the minor sections constitute rather high touch voltages for operator safety: they could be reduced by increasing the number of joints which would allow lessening the length of the major sections and consequently the distance between the earthing points.

The no-load steady-state regime #6 of Figure 3.36 has also been investigated: Figure 5.23 shows the voltage and current magnitudes along the cable line. The

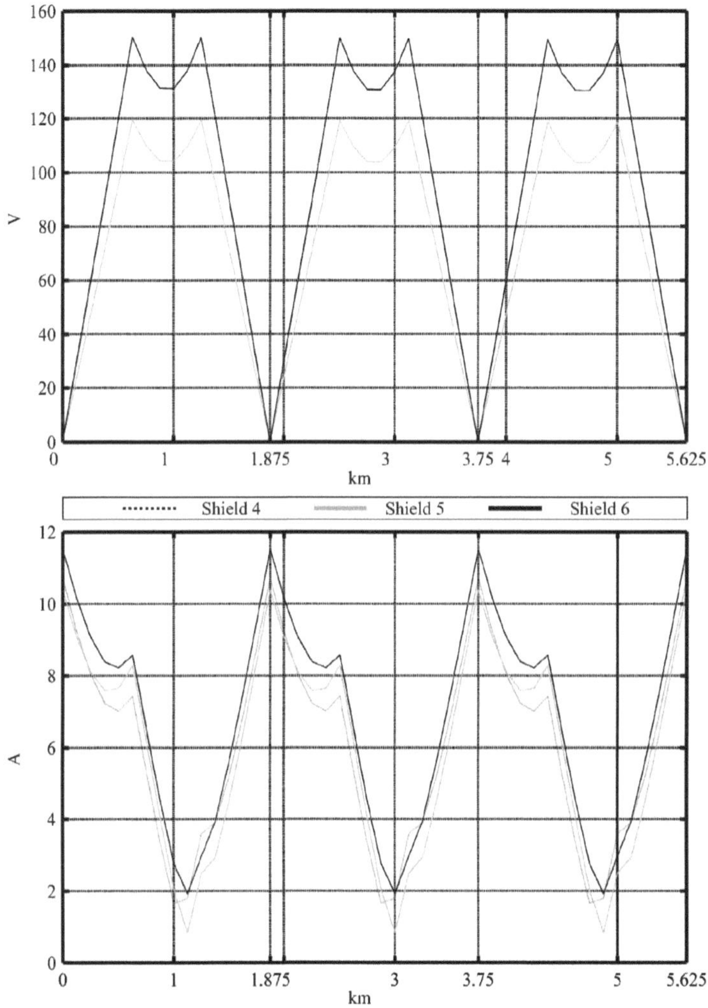

Figure 5.22 RMS sheath voltage and current magnitudes along the single-circuit cable #b, for the first three major sections ($\xi_{sh} = 0.608$ with 32 cross-bonding major sections)

behaviours are very similar to those shown in Figures 3.37 and 3.38. Moreover in Figure 5.24 the sheath voltage magnitudes are shown: it is extremely interesting to verify that the induced voltages in the sheaths are strictly depended upon the inducing phase currents.

Also the subtransient energization with and without lumped shunt compensation has been investigated with the multiconductor approach; in this case for the supply network modelling in subtransient regime (see in Section 5.5, Figure 5.9) the fol-

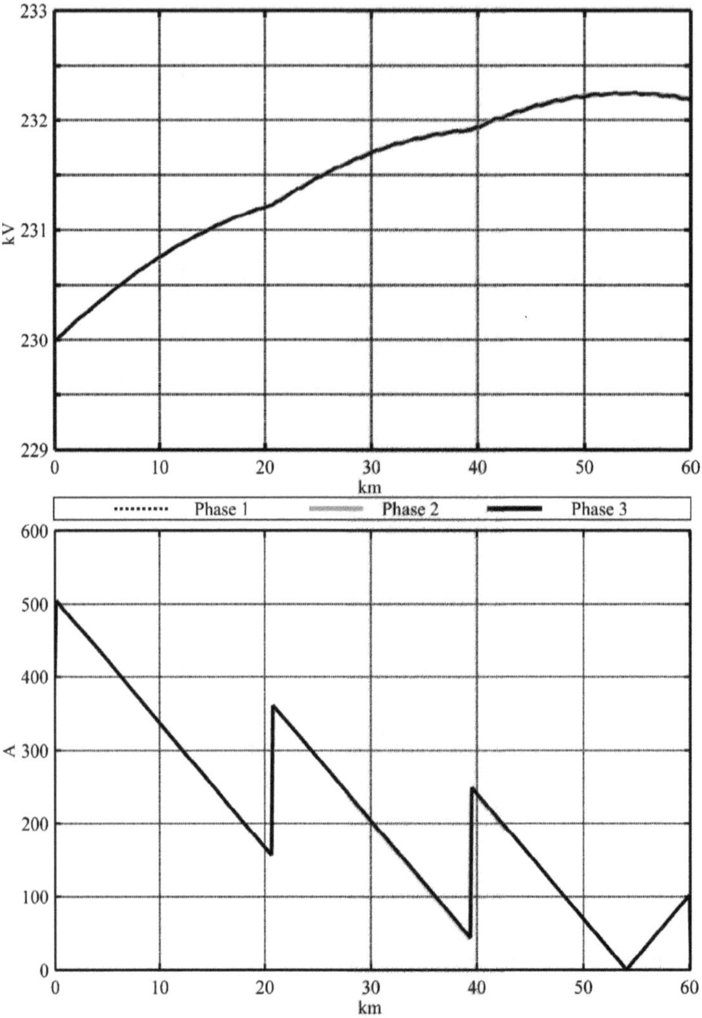

Figure 5.23 RMS phase voltage and current magnitudes along the single-circuit cable #b, for no-load energization (steady state) ($\xi_{sh} = 0.608$ with 32 cross-bonding major sections)

lowing values of the sequence impedances have been chosen: $\underline{z}''_{(+)} = \underline{z}_{(-)} = \mathrm{j}\,20\,\Omega$; $\underline{z}_0 = 5 + \mathrm{j}\,30\,\Omega$.

Table 5.2 shows the voltage magnitudes compared with those computed in Section 3.13: the accordance is fully satisfactory.

The matrix procedure here applied to the simple case of a single-circuit cable system can be enlarged to the study of more complex cable configurations (e.g. cross-bonded double-circuit in parallel electrical operation, with any configuration of supply phase sequence and any earthing techniques of the sheaths, [19]).

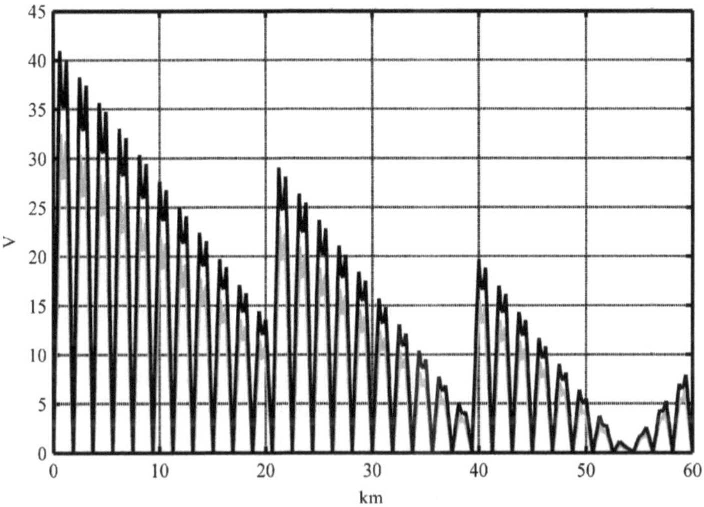

Figure 5.24 RMS sheath voltage magnitudes along the single-circuit cable #b, for no load energization ($\xi_{sh} = 0.608$ with 32 cross-bonding major sections)

Table 5.2 Subtransient energization voltage (kV) magnitudes $\underline{U}''_{0R}$ with and without shunt reactive compensation compared with the results derived in Chapter 3 (for the cable #b, $L = 60$ km, $\xi_{sh} = 0.608$)

		Multi-conductor analysis	Positive sequence analysis
With shunt reactive compensation	R	240.59	
	S	240.58	240.5
	T	240.60	
Without shunt reactive compensation	R	258.92	
	S	258.90	258.8
	T	258.93	

5.9.1 Comparisons with Other Methods

The comparison of the results obtained by means of the Carson–Clem’s formulae with those obtained by means of the complete Carson’s and Wedepohl’s gives a very good agreement: this means that, at power frequency, the use of Carson–Clem’s formulae is a powerful tool considering their elegant simplicity. Each case has been also compared with EMTP cable-constants [12, 20]: the percentage difference is a maximum of 1%, which no comparative figure of standard size can clearly highlight.

The presented multiconductor method is completely general and does not need any algebraic manipulation as in the previous contributions [21, 22] or loop current analysis [23].

5.10 Conclusions

The elegance and compactness of matrix notation applied to power system (which has been highly emphasized by Kron's Tensor Analysis [24]) find in multiconductor cell analysis a powerful and natural application.

The possibility of choosing in the theoretic settlement the simplified Carson–Clem's equations, the complete Carson's theory or Wedepohl's (which give the same results at power frequency) allows using the more suitable tool depending upon the cases.

Accuracy of the method has been verified through systematic comparison between multiconductor algorithms and the traditional software *EMTP cable-constants* (and *FEM* validation as well).

The multiconductor matrix analysis has demonstrated several advantages. The possibility of knowing the electric quantities along the line with a detail as better as shorter the cell length allows determining the behaviour of all system conductors including the earth; it constitutes a sure reference in order to compare the results of more simplified methods.

Other immediate applications of this multiconductor analysis could be:

- Gas Insulated Lines considering the phases plus the enclosures;
- Mixed line composed of cascade connection of overhead line (with one or two ground wires) and double or single-circuit UGC; and
- Cables in tunnel installation considering the steel reinforcement of the gallery as distributed earthing or other earthing conductors [25].

The multiconductor analysis allows:

- Detecting the exact repartition of currents in the phases and in the sheaths of any number of UGC circuits and with any bonding configuration (cross-bonding, single-point bonding, solid-bonding);
- Knowing the precise steady state electric behaviour of any component of the cable line (i.e. cross-bonding boxes);
- Considering possible phase transpositions;
- Including the presence of the metallic grid of the gallery hosting the transmission line (if tunnel installation);
- Studying the subtransient short-circuit regime and the voltages in any line section (consequently the touch-voltages);
- Studying accurately the power loss behaviour along the line and subsequently the temperature distribution and ampacity along the line with a suitable thermal model.

References

1. CCITT: Directives concerning the protection of telecommunication lines against harmful effects from electric power and electrified railway lines. Geneva, 1989
2. H.R.J. Klewe: Interference between power systems and telecommunication lines, E. Arnold London, 1958
3. Technical Brochure No. 95, Cigré Working Group 36.02: Guide on the influence of high voltage AC power systems on metallic pipelines, 1995
4. J. Arrillaga, C.P. Arnold: Computer analysis of power systems, John Wiley and Sons, England, 1990
5. N.D. Tleis: Power systems modelling and fault analysis, Elsevier Newnes, UK, 2008
6. J.R. Carson: Wave propagation in overhead wires with ground return, *Bell System Tech. Journal*, 5, 1926, pp. 539–554
7. F. Pollaczek: Über das Feld einer unendlich langen wechselstromdurchflossenen Einfachleitung, *Elektrische Nachrichten Technik*, Vol. 3, No. 9, 1926, pp. 339–360
8. A. Hochrainer: Symmetrische Komponenten in Drehstromsystemen, Springer, Berlin, 1957
9. L.M. Wedepohl, D.J. Wilcox: Transient analysis of underground power transmission systems, *Proc. IEE*, Vol. 120, No. 2, February 1973, pp. 253–260
10. IEC 60287: Electric cables – Calculation of the current rating, (in 8 parts: 1.1, 1.2, 1.3, 2.1, 2.2, 3.1, 3.2, 3.3)
11. R. Benato, R. Caldon: Distribution Line Carrier: Analysis Procedure and Applications to Distributed Generation, *IEEE Trans. on Power Delivery*, Vol. 22, Issue 1, January 2007, pp. 575–583
12. H.W. Dommel: Electromagnetic Transients Program Reference Manual (EMTP Theory Book), Prepared for Bonneville Power Administration, Portland, 1986
13. S.A. Schelkunhoff: The electromagnetic theory of coaxial transmission lines and cylindrical shields, *Bell System Technical Journal*, 13, 1934, pp. 532–579
14. C.L. Fortescue: Method of symmetrical co-ordinates applied to the solution of polyphase networks, Presented at the *34th Annual Convention of the American Institute of Electrical Engineers*, Atlantic City, N.J., June 28, 1918, pp. 1027–1140
15. C.A. Gross: Power System Analysis, John Wiley & Sons, second edition, 1986
16. R. Benato, R. Caldon, A. Paolucci: Automatic Procedure for Electromagnetic Interferences between Electric Systems of Different Nature, *Proc. of Italian Electrical and Electronics Association (AEI) Annual Meeting*, Vol. 3, 1997, pp. 19–26 (in Italian)
17. R. Benato, L. Fellin, D. Marzenta, A. Paolucci: Gas–insulated transmission lines: excellent performance and low environmental impact, *Proc. of EPETM*, May 12–18 2000, Napoli–Capri, Vol. 1, pp. 385–405
18. E. Peschke, E. von Olshausen: Cable systems for high and extra-high voltage, 1999, Publicis MDC Verlag, Germany
19. R. Benato: Multiconductor Analysis of Underground Power Transmission Systems: EHV AC Cables, *Electric Power System Research*, Vol. 79, Issue 1, January 2009, pp. 27–38
20. M. Kizilcay, M. Ermel, S. Demmig, H. Biewald: Modelling of a 400 kV XLPE cable system, Proceedings of EEUG 99, Italy, 1999, pp. 83–96
21. G. Bazzi: Cable sheath loss reduction by means of cross-bonding, *L'Elettrotecnica*, Vol. LVIII-N.4, April 1971, pp. 216–228 (in Italian)
22. C. Adamson, L.M. Wedepohl: Comparative steady-state performance of cross-bonded cable systems, *Proc. of IEE*, Vol. 115, No. 8, August 1968, pp. 1147–1156
23. R. Natarajan, T. Hunger: Investigations on three-phase current distribution in parallel cable conductors, *Electric Power Components and Systems*, Vol. 33, No. 7, July 2005, pp. 755–766
24. G. Kron: Tensor analysis of networks, MacDonald, London, 1965
25. R. Benato: Multiconductor Cell Analysis Of Power Cable Steady State, *Proc. of JICABLE 2007, Paper No. C.5.2.9.*, pp. 707–712, Versailles, France, 24–28 June

Chapter 6
A Comparative Procedure for AC OHL and UGC Overall Cost

6.1 Introduction

For a long period, starting from the XLPE EHV cable market injection, the high investment costs of EHV UGC (which can be increased by the shunt compensation), were often taken as a sure cause to prefer a priori an overhead line in the network widening or reinforcing, without devoting the right consideration to the (sensibly different from cable to overhead lines) burdens deriving from territory impact and from the energy losses in the whole line lifetime.

The importance of these costs, at first disregarded, has become evident successively [1] due to both the increased ecological-economical needs of the territory and the quality of power grid energy efficiency: consequently suitable criteria were introduced in order to evaluate the costs regarding both the territory impact and the energy losses. These ones (burdens of territory impact and energy losses) must be added (together with others of less relevance e.g. O&M, dismantling ones) to the mere investment cost of a new line so to have a more meaningful "*overall cost*" for a right comparative analysis. In particular, the financial burdens on the territory (both private and public property) caused by the line along the whole route have generally a not negligible economic weight: they must be computed and introduced (as compensation costs) in a complete evaluation of an "*overall cost*" due to the erection of a new line.

This methodology, which takes into account the technical, environmental and social aspects involved in the overall economy of a new line, seems to be in whole accordance with the aims of the "Environmental Impact Assessment" recommended by the European community since 1985 [2].

In this chapter the following costs will be considered:

- capital costs (I);
- loss energy costs (E);
- burden on territory (T);
- a short account on the visual impact;
- operating and maintenance (O&M) costs (OM);

- dismantling costs (D);
- UGC shunt compensation investment cost $(\Delta I)_{Sh}$; and
- overall cost (Σ).

The costs deriving from the forced outages and the failure reparation are not considered as it will be explained in Section 6.7.

The comparison procedure proposed in this chapter is limited to the power frequency technologies but in the technical literature there are similar approaches [3, 4] to include DC transmission.

In the following, in order to substantiate the procedure, exemplificative values of the parameters are chosen: even if they may vary year by year and country by country (especially those of economic nature), the methodology and approach remain always valid.

6.2 OHL and UGC in the Comparative Procedure

Figures 6.1 and 6.2 show the configurations of OHL (#a1) and UGC (2#c1) respectively (both at 380 kV, 50 Hz), which will be considered in the description of the comparative procedure.

Figure 6.1 shows a typical tower used (with different suitable heights) in the EHV Italian grid, and reports the necessary data to calculate, by means of the usual formulae, the positive-sequence line parameters by hypothesizing phase transpositions and neglecting the earth wires (see Table 6.1); also in the cases of untransposed lines, the errors are slight for the purposes of this chapter.

The line configuration is important also for computations of the magnetic and electric fields and the width of the right-of-way which results, after Italian laws [5–7], quite considerable (see Section 6.5) so to weight sensibly on the territory economy and therefore in the results of the total cost comparisons.

A traditional tower only is here considered even if overhead lines could be erected nowadays with more innovative design (compact towers, Foster type, etc.).

The per unit length resistance of OHL has been computed at 75 °C, conductor temperature when operated at the thermal limit after the Italian Standard CEI 11-60 [8]. It is worth underlining that the operation of an untransposed OHL at thermal limit current increases the voltage unbalance factor [9] and it is advisable only for sufficiently short lines (as those hereby considered).

For technical evaluations at power frequency, the OHL conductance (g_1 or g_2 of Table 6.1) can be usually neglected, but not for economical evaluations: corona losses and the insulators leakage currents depend strongly [10, 11] upon the weather conditions which must be accurately evaluated chiefly where the rainy days often occur during the year.

Figure 6.2 shows the underground installation of the proposed double-circuit 2#c1: its fundamental data are reported in Tables 6.2 and 6.3.

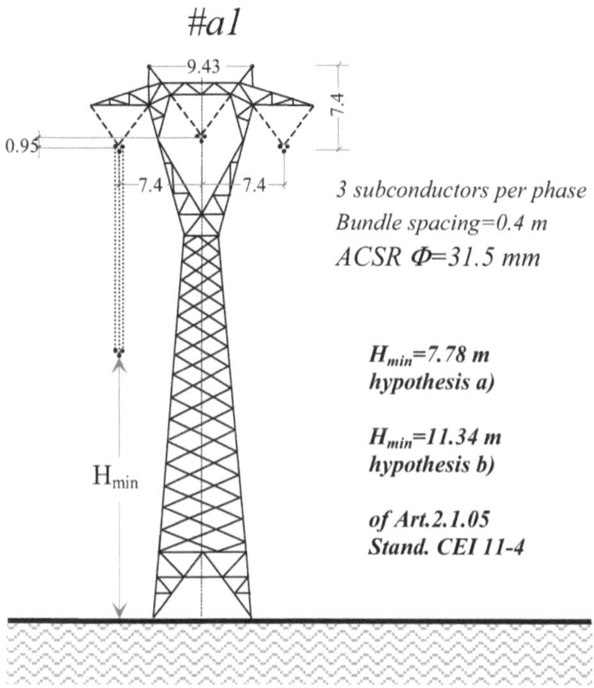

Figure 6.1 Typical configuration of an OHL in the 380 kV Italian grid

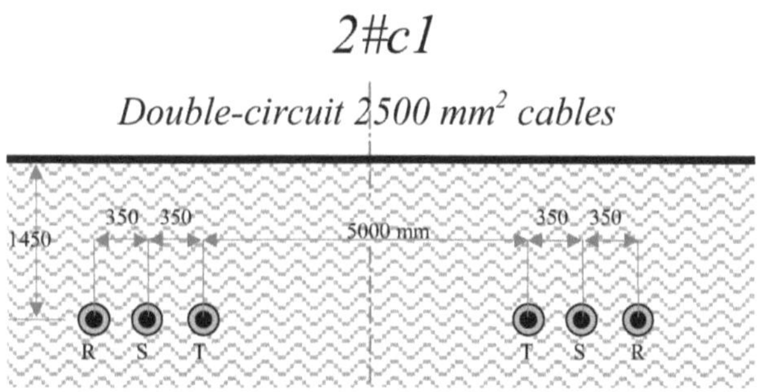

Figure 6.2 Installation of 380 kV cross-bonded, XLPE UGC (2 single-circuits #c1 in electrical parallel operation)

Table 6.3 reports some values of the ampacity depending upon the different soil thermal resistivities ρ_{th} which are within a wide range foreseen in Clause 3.2.2 of IEC 60287-3-1 [12].

Owing to the great spacing (5 m), the two single-circuits #c1 are thermally and electrically almost independent so that, for double-circuit 2#c1, the ampacities are twice the I_a values in Table 6.3 and the electrical model can be performed after Section 2.7 by means of the single-circuit #c1 parameters.

Table 6.1 Positive-sequence parameters of transposed OHL #a1 (ignoring the earth wires)

Overhead				#a1
Nominal voltage				380 kV
Phase bundle				3 sub-cond. ACSR $\Phi = 31.5$ mm spacing = 0.4 m
Resistance at 75 °C (50 Hz)		r	mΩ/km	23.10
Inductance		ℓ	mH/km	0.858
Capacitance		c	μF/km	0.0133
Shunt leakage (50 Hz)	Fair weather Rainy weather	g_1 g_2	nS/km	10 80
Ampacity (Zone A-Italy) after CEI 11-60	Cold Months (Oct.–Apr.)	I_a	A	2955
	Warm Months (May–Sep.)			2220

Table 6.2 Cable geometrical data

Single-core cable		#c1
Nominal voltage		380 kV
Cross-sectional area	mm^2	2500 Cu
Diameter on conductor (Milliken)	mm	63.4
Diameter on XLPE insulation	mm	119.9
Diameter on metallic screen	mm	130.1
Cross-section of screen	mm^2	≈ 500 Al
Diameter on PE coating	mm	141.7
Mass	kg/m	37

Table 6.3 Positive-sequence parameters of cross-bonded UGC (single-circuit)

UGC				#c1
Nominal voltage				380 kV
Cross-section			mm^2	2500 Cu
Resistance at 90 °C (50 Hz)		r	mΩ/km	13.3
Inductance		ℓ	mH/km	0.576
Shunt leakance (50 Hz) with $\tan\delta = 0.0007$		g	nS/km	51.5
Capacitance with $\varepsilon_r = 2.3$		c	μF/km	0.234
Steady state ampacity I_a	with $\rho_{th} = 0.8$ K m/W (case α_1)	I_a	A	1975
	with $\rho_{th} = 1.0$ K m/W (case α_2)	I_a	A	1788
$\vartheta = 90$ °C (XLPE);	with $\rho_{th} = 1.2$ K m/W (case α_3)	I_a	A	1676
$\vartheta_a = 20$ °C	with $\rho_{th} = 1.4$ K m/W (case α_4)	I_a	A	1566
	with $\rho_{th} = 2.0$ K m/W (case α_5)	I_a	A	1327

ρ_{th} = soil thermal resistivity

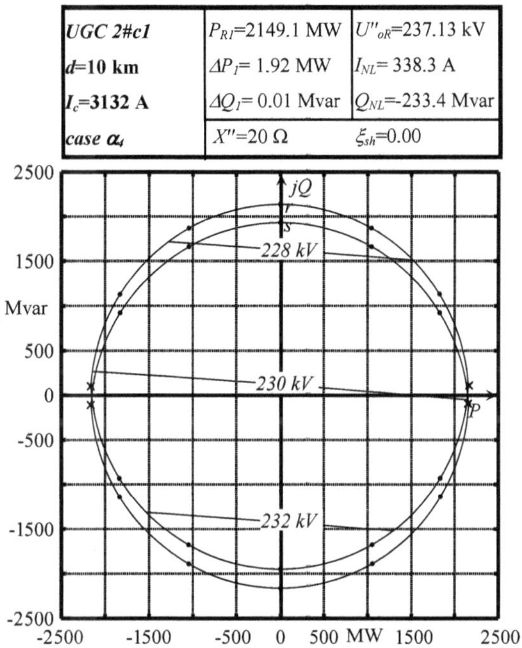

Figure 6.3 Capability charts of 2#c1 (case α_4)

It is absolutely worth outlining that the comparison here proposed (#a1 versus 2#c1) is a simple example in order to apply the procedure and does not imply a univocal correspondence between OHL and UGC.[1]

It rises not only from the great variety of the possible installations within the different national rules, but also from the environmental conditions, different zone by zone for each country, from the evolution of the different technologies and from the criteria more or less conservative used for planning a line.

In order to confirm a possible operating equivalence of 2#c1 against of #a1 the corresponding capability charts must be compared so that it is necessary also for 2#c1 to fix the achievable level I_c: it depends not only upon the electrical and thermal parameters of the cable itself but also upon the thermal parameters of the soil (see Table 6.3) where it is laid. It is well known that the different thermal resistivities of the soil (chiefly depending upon the moisture content) may strongly vary along the route [13] and, unless an extensive measurement campaign is performed, they cannot be known with great detail. It is also worth remembering the uncertainties of the forecast of the soil temperatures, both for the climate behaviours and for the possible presence of external heat sources nearby the route [14].

Once fixed, in a more or less conservative way, the I_c level compatible with the environmental conditions (e.g. case α_4) and the length of the line ($d = 10$ km as #a1), it becomes possible to compute the capability charts of 2#c1 (see Figure 6.3

[1] Clause 5 of IEC 60287-3-1 gives the purchaser a useful guide in order select the appropriate type of cable together with the manufacturer.

with $I_c = 2 \cdot 1566 = 3132$ A) beyond that of #a1 (see Figure 6.4): by means of these charts, it can be verified that the two lines are able to face satisfyingly the same operating conditions (see load duration curves in Sections 6.4 and 6.10).

Figure 6.5 clearly highlights the reduction of the receiving area for #a1 in "warm period" which must be well considered in the operation during the year: it also shows the possible advantages of 2#c1 in more thermally favourable soils (cases α_1–α_3).

Particular attention must be paid to the different foreseeable levels of availability: since the great repairing times of a UGC against those of OHL, the choice of a double circuit UGC (as here considered 2#c1), seems to be suitable for emergency operation when one of the two circuits is unavailable (e.g. see Section 6.10.2). To this aim, it becomes absolutely necessary to consider also the capability charts of #c1.

6.3 The Capital Costs of OHL and UGC

In a first approach, the investment costs of the terminal stations are approximately equal for OHL and UGC (single or double cable circuits) and so disregarded in the comparison; in a detailed analysis, they could be taken into account.

Beyond the purchasing and installation costs, the capital cost includes the cost of other additional works for the lines (e.g. excavation, ducts, tower foundations, wayleave).

By defining

- a = capital cost of OHL #a1 per kilometre (M€/km),
- c = capital cost of UGC 2#c1 per kilometre (M€/km),

the following values will be hypothesized, in order to develop the case studies of Sections 6.11 and 6.12,

$$a = 0.35 \text{ [M€/km]} \; ; \quad \frac{c}{a} = 10 \, .$$

It is worth noting that the values are merely indicative since they can strongly vary with market conditions.

6.4 Energy Losses and Their Actual Cost

The economical assessment of the energy losses during the operational life of a transmission line plays a significant role in the overall cost evaluation.

The energy losses which have a meaningful economic weight can be computed by the steady state power flow studies: the power flow diagrams (and hence the power losses) are strictly linked to the line typology inside the electric network structure: it is worth remembering for example, in EHV, cross-border interconnections (complex power in the first and third quadrant), connections between power plant and grid [15–18] (complex power in the first quadrant) or lines inside the meshed transmission network.

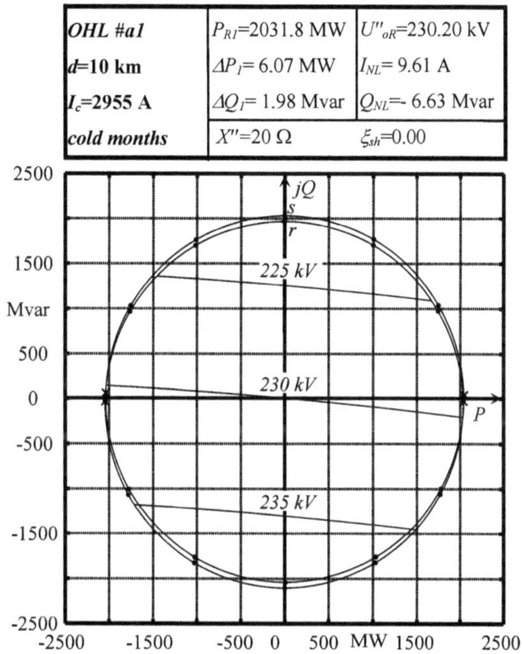

Figure 6.4 Capability charts of #a1 (cold months)

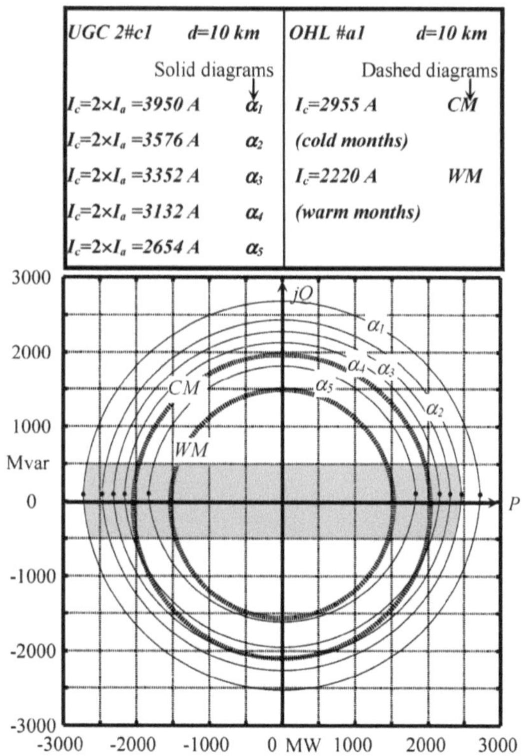

Figure 6.5 Capability charts (only receiving areas) of 2#c1 for the different ampacities I_a of Table 6.3 and those of #a1; (in *grey area* $|Q_R| \leq 500$ Mvar)

With regard to the latter, the power flow diagrams present generally great fluctuations both on a daily and monthly basis due to the variations in the load set. For the other two line typologies, the power flows can present a constant behaviour on a long term (e.g. connections for base service power plants, cross-border interconnections) but also strongly variable behaviours (e.g. connections for regulation service power plants).

In the present case studies (see Section 6.10), any power flow diagram (with the ratio between reactive and active power $Q_R/P_R = 0.2$) has been represented by means a typical annual duration curve, which efficaciously reports the magnitudes of the receiving power P_R surveyed each hour of the year (see, for example, Figure 6.6 in the hypothesis of perfect availability of 8760 h).

In order to simplify the computation procedure, it is preferable to approximate the continuous curve of P_R with the step curve (see Figure 6.6), where $\Delta\tau = 1000$ h (but the last one $\Delta\tau = 760$ h), so that for each time step $\Delta\tau$ it can be assumed a constant complex power $\underline{S}_\tau = P_\tau(1 + j0.2)$. Since the chosen step diagram must be the same for both the line typologies, a greater accuracy (e.g. with $\Delta\tau = 500$ h) is not necessary.

By using the Osanna's method suitably implemented, it is immediate to compute the power frequency regime of the line due to each complex power $\underline{S}_\tau$ at receiving end and hence the energy losses in the considered time step.

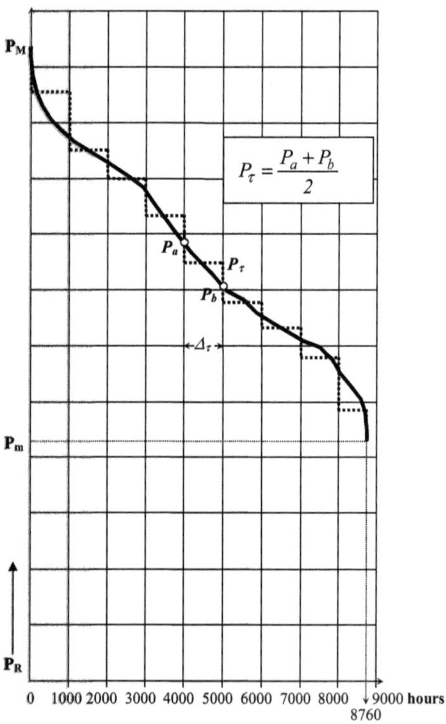

Figure 6.6 A typical year duration curve of receiving active power P_R (*solid line*) and approximated step diagram (*dashed line*)

In such a way, the different foreseeable annual energy losses *AEL* for OHL and UGC (with the same line length d) and subjected to the same duration curve of demand power can be evaluated.

In the following computations, during the n years of the UGC and OHL life, the duration curve, the annual discount rate and the loss energy specific cost are assumed unchanged: by giving the energy losses a specific cost,

- **Loss energy specific cost** $= Lec$, e.g. $Lec = 40$ [€/MW h];

the annual loss cost is given by

$$ALC = AEL \cdot Lec\ ;$$

hence, the corresponding actual costs V (V_{OHL} and V_{UGC}) can be computed by means of

$$V = \frac{(1+i)^n - 1}{i(1+i)^n} \cdot ALC\ . \tag{6.1}$$

By assuming:

- **Line lifetime** $= n = 35$ [years];
- **Real rate of annual interest (discount rate)** $= i = 0.025$ ($\Rightarrow$ 2.5%),

it yields:

$$V = 23.15 \cdot ALC\ .$$

In any case, all power line life long, it appears problematic to foresee:

a) the behaviour of power flow diagrams, generally heavier as years go;
b) the energy loss specific costs depending upon the market evolution;
c) the different annual discount rates for the financial calculation of actual costs of energy losses.

However, by implementing some algorithms of financial mathematics in year intervals with constant parameters, it is possible to develop a numerical program evaluating the sensibility of (E) and (Σ) to the foreseeable variations of a), b) and c).[2]

The values reported in Tables 6.4 and 6.5 regard idealized cases (line length of 1 km along which the voltage and current remain practically constant) but allow the reader to realize the different contribution of P_{J} and P_{g} to the total power losses.

In the OHL the losses due to the insulators and to the corona-effect increase severely under foul weather conditions [10, 11] which in some countries may occur in several days of the year.

It is of note that the Joule losses have always a prevailing weight both for OHL and UGC and that OHL Joule losses are always greater than those in UGC owing to the major longitudinal resistance ($r_{\#a1}/r_{2\#c1} \cong 3.5$). It yields also annual energy losses *AEL*, computed on the basis of the same duration curves, much higher than those of UGC and subsequently $V_{\mathrm{OHL}} \gg V_{\mathrm{UGC}}$.

[2] IEC 60287-3-2 gives useful comments on these economic aspects.

Table 6.4 Power losses [kW/km] of OHL #a1 and UGC 2#c1 with line length = 1 km, I = 2955 A at U_n = 380 kV

		I = 2955 A	
#a1	Power losses kW/km	2#c1 $\xi_{sh} = 0$	Power losses kW/km
P_J	605.10	P_J	174.2
P_{g1}	1.45	P_g	15
P_{g2}	11.60		
$P_J + P_{g1}$	606.6	$P_J + P_g$	189.2
$P_J + P_{g2}$	616.7		

P_J = Joule losses (UGC and OHL)
P_{g1} = insulator and corona losses in OHL (fair weather)
P_{g2} = insulator and corona losses in OHL (rainy weather)
P_g = dielectric losses in UGC

Table 6.5 Power losses [kW/km] of OHL #a1 and UGC 2#c1 with line length = 1 km, I = 1500 A at U_n = 380 kV

		I = 1500 A	
#a1	Power losses kW/km	2#c1 $\xi_{sh} = 0$	Power losses kW/km
P_J	156.00	P_J	44.90
P_{g1}	1.45	P_g	15.00
P_{g2}	11.60		
$P_J + P_{g1}$	157.5	$P_J + P_g$	59.90
$P_J + P_{g2}$	167.60		

P_J = Joule losses (UGC and OHL)
P_{g1} = insulator and corona losses in OHL (fair weather)
P_{g2} = insulator and corona losses in OHL (rainy weather)
P_g = dielectric losses in UGC

Therefore for each kilometre of line length #a1 the discounted energy loss cost is given by:

$$(\mathsf{E})_a = \left(\frac{V_{OHL}}{d}\right) \quad [\text{M€/km}] ,$$

and analogously for UGC 2#c1:

$$(\mathsf{E})_c = \left(\frac{V_{UGC}}{d}\right) \quad [\text{M€/km}] .$$

6.5 The Burden on Territory

Nowadays, notwithstanding the numerous researches on magnetic induction exposure [19–21], there is not yet a worldwide recognized magnetic limit value even if the Guidelines of International Commission on Non-Ionising Radiation Protection (ICNIRP) recommends some reference levels for the general public exposure to time-varying electric and magnetic fields [22].

In Italy for instance, there is a general discipline [5, 6] devoted to the population protection from electromagnetic field exposure at 50 Hz. There are different limit values B [μT] for short-term and long-term exposure; the short-term exposure is 100 μT in accordance (at 50 Hz) with ICNIRP [22]. With regard to long-term exposure, for existing lines there is a value of attention of 10 μT, whereas for new lines, a "target of quality" of 3 μT (RMS value).

It is worth remembering that in Switzerland for new installations the average magnetic field long term exposure must not exceed 1 μT [23].

The Italian Law compels to calculate the magnetic fields generated by power lines at the ampacity regimes and to individuate [7, 24], along the route, the width F of the "corridor" where $B \geq 3$ μT; in the "corridor" any extended residence (> 4 h a day) is interdicted.

The width F depends upon the phase currents and the line arrangement; for the computation of magnetic fields there are several guides [25–27].

Figure 6.7 shows the magnetic induction levels generated, in a plane perpendicular to the line, by the double-circuit UGC 2#c1 with $I_{\text{phase}} = 2955$ A considering the sheath currents zeroed by the cross-bonding. In any case, the multiconductor analysis (see Chapter 5) would allow a more detailed analysis. It is of note that, for UGC, the spacing reduction and/or metallic screening decrease F_c but imply a more dissipative installation and hence a lower ampacity.

Figure 6.8 gives an immediate view of the magnetic induction levels generated by the OHL #a1 with $I_{\text{phase}} = 2955$ A and span attachments at 32–33 m, neglecting the current induced in the earth wires. In many cases, instead of computing the different corridor widths on the ground along the span, it is allowable [7, 24] to assume conservatively the vertical projection of F_a on the ground itself. Also for OHL the use of mitigation screening circuits is possible but they are extremely onerous [28]. A very slight screening effect is indirectly (see $F_a = 101.3$ m in Figure 6.9) given by the earth wires, whose currents have been computed by means of multiconductor algorithms.

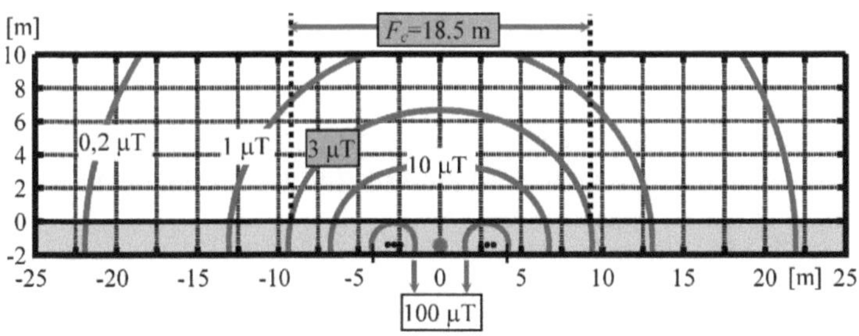

Figure 6.7 Magnetic induction levels for the double-circuit cross-bonded UGC (2#c1, $I_{\text{phase}} =$ 2955 A) and width F_c of corridor

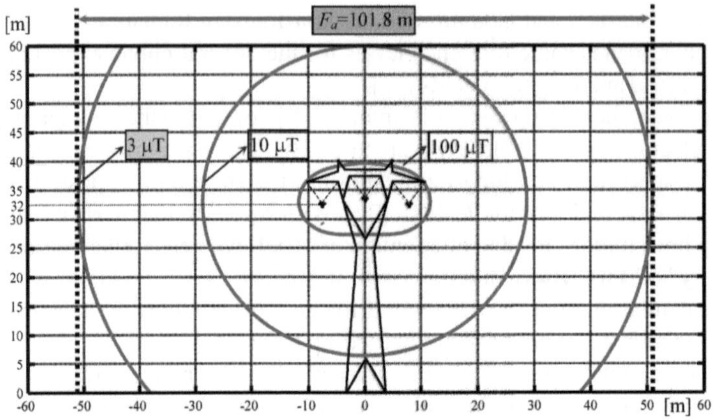

Figure 6.8 Magnetic induction levels for the OHL with $I_{\text{phase}} = 2955$ A (span attachments at 32 and 33 m) neglecting the presence of earth wires

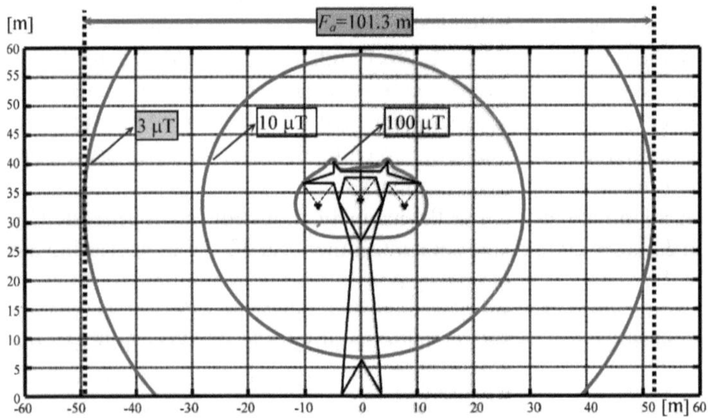

Figure 6.9 Magnetic induction levels for the OHL with $I_{\text{phase}} = 2955$ A considering the presence of currents of earth wires by means of multiconductor algorithms

The ratio $F_a/F_c = 5.5$ is a first impressive index of the different territory impacts.

Since in the area where $B \geq 3\,\mu\text{T}$ any extended residential building is not permitted, it appears necessary to evaluate the loss of value inflicted on the territory itself by the new line. To this aim, it is proposed to presume for it an "average building parameter *ed*" similar to that of adjacent areas.

The value of *ed* $[\text{m}^3/\text{m}^2]$ is highly variable in the territory and means the average ratio, in a given area, between the developed building volume and the surface of area itself; it ranges between 3–4 (or much greater) in urban area and 0.8–1 in suburban

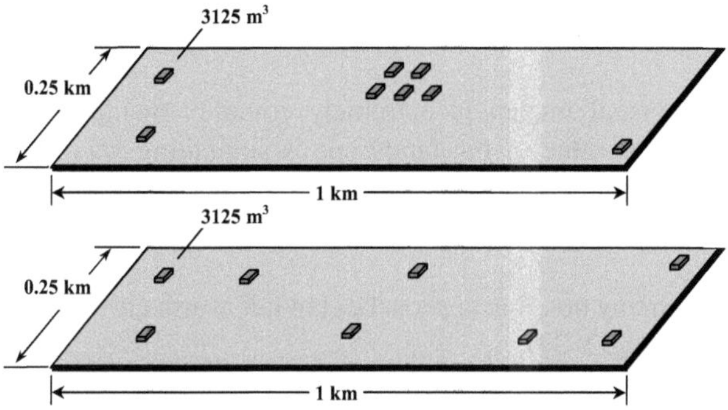

Figure 6.10 Visual impression of two areas with the same average building parameter $ed = 0.1$ $[\text{m}^3/\text{m}^2]$

area and can lessen below 0.1 in rural areas. For the sake of a visual impression, see Figure 6.10 where two areas have the same average building parameter $ed = 0.1$ $[\text{m}^3/\text{m}^2]$.

As a first approach, it can be considered that every square-meter of land located in the "right-of-way" (corridor of width F) would lose a value:

$$w_x = k \cdot ed \quad [€/\text{m}^2] , \tag{6.2}$$

where k depends strongly upon the property market. A merely indicative value could be set equal to $k = 400\,€/\text{m}^3$ so that when $ed = 0$–$0.1\,[\text{m}^3/\text{m}^2]$.

$$w_x = 0\text{–}40 \quad [€/\text{m}^2] . \tag{6.3}$$

In fact, even if $ed = 0$, a rural area inside the right-of-way has nevertheless a real damage since it will be no more possible to erect any farmer residence.

In the following comparisons, the reader could ascertain that the variations of w_x will have a great influence.

In conclusion, depending upon the foreseeable value of w_x, for each kilometre of line #a1, it can be supposed, on the territory extent $F_a \cdot 1000\,[\text{m}^2]$, the damage:

$$(\text{T})_a = F_a \cdot 10^3 \cdot w_x\ [€/\text{km}] \quad \Rightarrow \quad F_a \cdot 10^{-3} \cdot w_x\ [\text{M}€/\text{km}] = 0.1 \cdot w_x$$
$$(\text{with } F_a = 100\,\text{m}) ;$$

analogously for UGC 2#c1:

$$(\text{T})_c = F_c \cdot 10^3 \cdot w_x\ [€/\text{km}] \quad \Rightarrow \quad F_c \cdot 10^{-3} \cdot w_x\ [\text{M}€/\text{km}] = 0.018 \cdot w_x$$
$$(\text{with } F_c = 18\,\text{m}) .$$

6.6 The Visual Impact

The economic evaluation of visual impact is extremely complex owing to its strongly subjective nature (as the value of the landscape is something very specific and a function of local views and preferences). Notwithstanding the ambiguity, when a new line must be installed, this aspect could be evaluated as in Chapter 15 of [29] and in [30, 31].

The economic theory offers many possible approaches but it is worth citing those based upon contingent evaluation [32–34].

Sometimes, the privilege of UGC on the visual impact is so evident that any economical comparison is not worth proposing.

6.7 Operation and Maintenance (O&M) Costs

The planned operation and maintenance of a line, during its life, implies some costs, which must be considered in the overall cost analysis.

The annual cost $AC_{\mathrm{O\&M}}$ of O&M is a percentage of investment cost.

Of course, the annual costs $AC_{\mathrm{O\&M}}$ must be discounted by means of

$$V_{\mathrm{O\&M}} = \frac{(1+i)^n - 1}{i(1+i)^n} \cdot AC_{\mathrm{O\&M}} \tag{6.4}$$

analogously to (6.1).

For an OHL, the maintenance costs are between 0.7 and 1% a year (with respect to the investment cost) and depend upon the weather and environmental conditions. The OHL operation cost ranges between 0.8 and 1%. The OHL O&M (flat installation with low salt pollution) can range between 1.5 and 2%. These values (see Table 6.6) must be considered as an average indication and can increase in case of extraordinary environmental occurrences.

With regard to UGC, once installed, they do not need particular maintenance due to the absence of atmospheric external situations.

The UGC maintenance can be evaluated as 0.1% of the capital cost (see Table 6.6).

The procedure does not take into account the different failure repairing times of OHL and UGC and their influences on the system costs.

Table 6.6 O&M annual costs as a percentage of investment costs

	OHL	UGC
Operation	0.8–1.0%	0.1–0.3%
Maintenance	0.7–1.0%	0.1%
Operation and maintenance (O&M)	1.5–2.0%	0.2–0.4%

To this regard, [35] and [36] give some values for the failure rates of EHV OHL and XLPE-insulated UGC respectively. An analogous issue is developed in [37] for GIL technology.

The procedure does not take into account the costs due to undelivered energy nor the congestion and re-dispatching costs arising from the differences in UGC and OHL failure repair times. Their burdens depend upon the possibilities and adaptability offered by the network meshing and consequently they are evaluable only by knowing the surrounding network; a more detailed analysis regarding the two line typologies would be misleading.

In conclusion, for OHL #a1, chosen O&M = 2%, the discounted value is given by:

$$(\mathsf{OM})_a = 0.02 \cdot 0.335 \cdot 23.15 = 0.16 \quad [\text{M€/km}] \ ,$$

and for 2#c1, chosen O&M = 0.4% it has:

$$(\mathsf{OM})_c = 0.004 \cdot 3.5 \cdot 23.15 = 0.32 \quad [\text{M€/km}] \ .$$

6.8 Dismantling or Decommissioning Cost

A comprehensive analysis of a transmission line cannot disregard the end of life i.e. the dismantling phase of it. This operation foresees some costs in order to restore the place previously occupied by the line itself: it is evident that the considerable delay with respect to the investment date gives a consequent low discounted value.

A possible order of magnitude is evaluated in [15] which gives a discounted value of the OHL dismantling cost:

$$(\mathsf{D})_a = \frac{17.500}{(1+i)^n} = 0.007 \ [\text{M€/km}] \quad (i = 2.5\% \text{ and } n = 35) \ .$$

For UGC 2#c1, the same reference [15] reports a value of 5% of the capital cost so that:

$$(\mathsf{D})_c = \frac{0.05 \cdot 3.5}{(1+i)^n} = 0.037 \ [\text{M€/km}] \quad (i = 2.5\% \text{ and } n = 35) \ .$$

6.9 The Cost of UGC Shunt Reactive Compensation

The circuital modelling of UGC shunt compensation (uniformly distributed or lumped) has already been throughout described in Chapters 3 and 4; in this paragraph there are some suggestions to introduce the relevant costs (capital costs and energy losses).

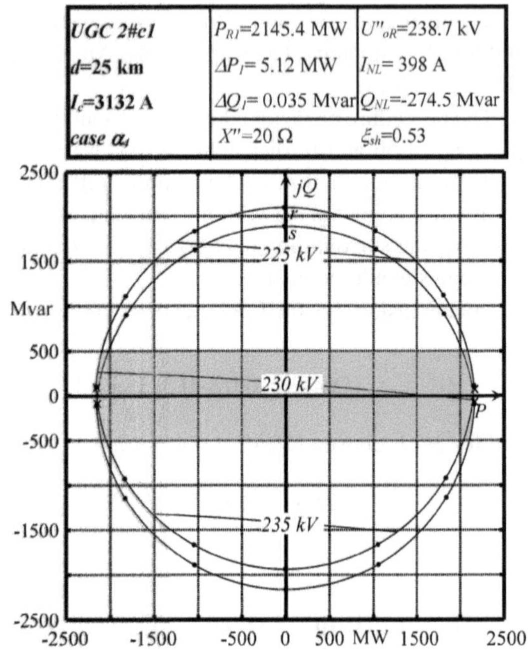

Figure 6.11 Capability charts of UGC 2#c1 by assuming an idealized distributed compensation

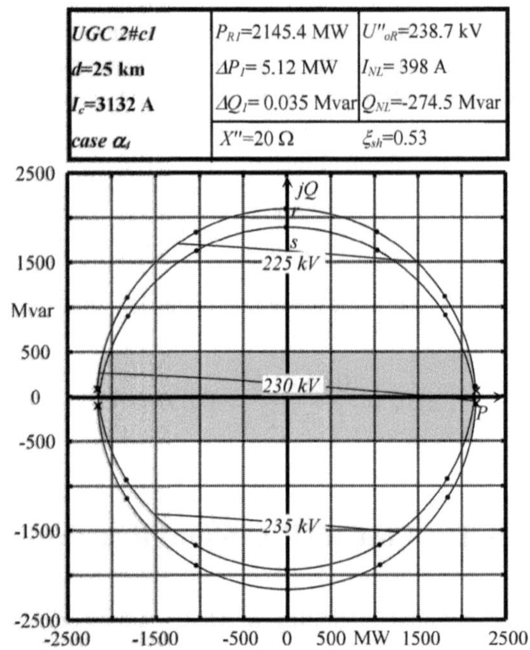

Figure 6.12 Capability charts of UGC 2#c1 by assuming lumped compensation at both ends

In order to consider a real case, let us consider the two lines #a1 and 2#c1 (with abovementioned parameters and ampacities) with the same length $d = 25\,\text{km}$. For such a length, the UGC 2#c1 requires the compensation degree $\xi_{\text{sh}} = 0.53$ which gives the capability charts of Figure 6.11 (uniformly distributed compensation) and of Figure 6.12 (lumped compensation at both ends considering also the reactor power losses).

Figure 6.13 shows the capability charts of the uncompensated #a1.

Therefore, it is possible to verify that the capability chart of Figure 6.12 (lumped compensation) gives a really good agreement with that of Figure 6.11 (distributed compensation) and presents conservative margins (grey zones in the Figures 6.11 and 6.12 and better levels U_{0R}) compared with the capability chart (OHL #a1) of Figure 6.13, so that, in abovementioned grey zones, any complex power $\underline{S}_R$ compatible with #a1 are also compatible with 2#c1.

The computation of annual energy losses of 2#c1 (followed by the discounted cost evaluation) must be performed by applying Ossanna's method (as already suggested in Section 6.4) on the matricial model of the lumped compensated circuit (see Chapter 2).

With regard to the investment costs, the increase Δ_{sh} must be added to the UGC investment cost (see Section 6.3): it is due to the costs of i) purchase and installation of the shunt reactors at both ends (usually three-phase with unchained magnetic flux), of ii) the disconnecting and breaking devices and the cost of iii) the territory occupation.

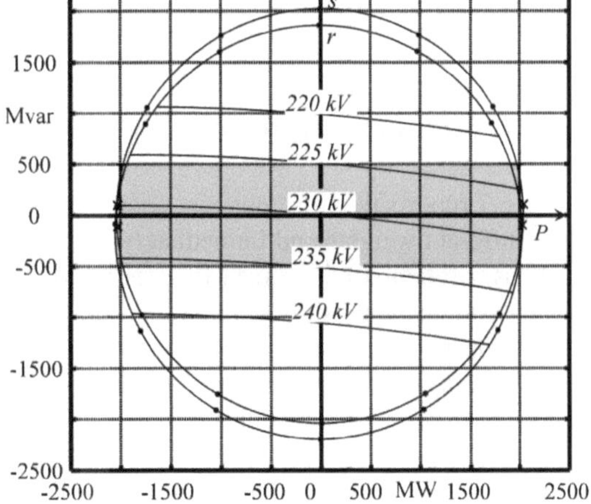

Figure 6.13 Capability charts of uncompensated OHL #a1

It is worth remembering that, in order to continue the operation also in emergency situation with a unique UGC circuit, it becomes necessary to foresee suitable taps on the three-phase reactors to reduce the compensation power, so avoiding the more expensive solution with two shunt reactors.

6.10 Two Case Studies: #a1 vs. 2#c1 with $d = 10\,\text{km}$

In the first case study, the duration curve of Figure 6.14a is assumed: it is suitable for a line belonging to the meshed network; in the second one the duration curve of Figure 6.14b, typical for a base power plant line, is considered.

As already stated in Section 6.4, the receiving-end complex power is $\underline{S}_R = P_R(1 + \text{j} \cdot 0.2)$; it can be ascertained that also when $P_R = 1800\,\text{MW}$, the corresponding $\underline{S}_R$ is compatible with the capability charts of Figures 6.3 and 6.4 (if, for OHL #a1, the regimes at $P_R = 1800\,\text{MW}$ are hypothesized in the "cold months").

The energy losses and the corresponding discounted costs have been computed in accordance with Section 6.4; for the greater losses in #a1 due to foul weather (see Table 6.1), the hypothesis of 90 days a year with four rainy hours is assumed.

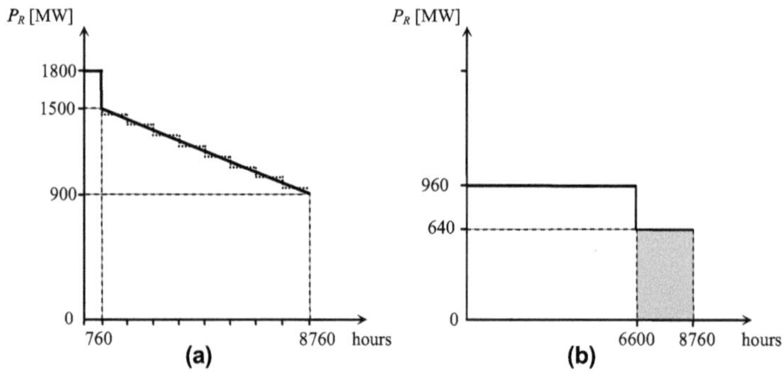

Figure 6.14a,b Schematic year duration curves

6.10.1 First Case Study with Duration Curve of Figure 6.14a

Table 6.7 reports the different cost components so that it is immediate to ascertain their different weights and immediately suggests to plot the overall costs $(\Sigma)_a$ and $(\Sigma)_c$ as a function of w_x (see Figures 6.15). It is worth noting that the term (T) comes from the results of Section 6.5.

Obviously, the abscissa $w_x = 24.3\,€/\text{m}^2$ of the intersection point (point of economical indifference) represents the value for which $(\Sigma)_a = (\Sigma)_c$ and clearly delimits the intervals of convenience.

It easy to verify that generally OHLs, having low investment costs but wide corridors F_a, are convenient in lands potentially suitable for building with moderate w_x.

Table 6.7 Composition of the overall cost [M€/km]

	#a1	2#c1
(I)	0.35	3.5
(E)	2.05	0.70
(OM)	0.16	0.32
(D)	0.007	0.037
(T)	$0.1 \cdot w_x$	$0.018 \cdot w_x$
(Σ) = (I) + (E) + (OM) + (D) + (T)	$2.57 + 0.1 \cdot w_x$	$4.55 + 0.018 \cdot w_x$

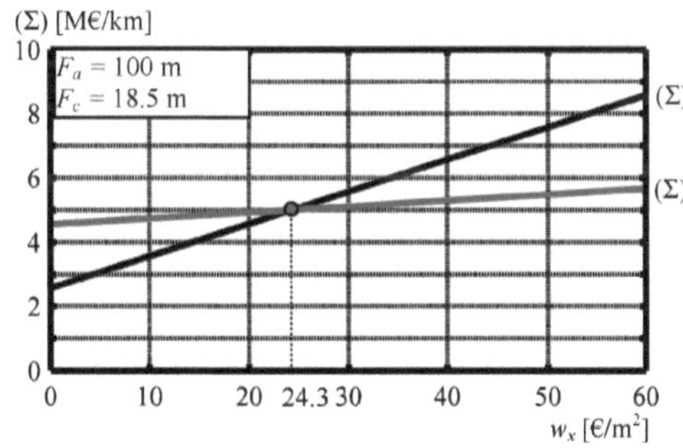

Figure 6.15 Overall costs per kilometre (with target of quality = 3 μT)

6.10.2 Second Case Study with Duration Curve of Figure 6.14b

The duration curve of Figure 6.14b can match the annual diagram of the complex power $S_R = P_R \mathrm{j}\, 0.2$) which must be delivered at node R by a thermoelectric power plant (three groups rated 320 MW) located at node S ($d = 10\,\text{km}$); the reduced value 640 MW for 2160 hours corresponds to the unavailability of each group for the planned maintenance of 30 days.

Table 6.8 shows the composition of overall costs where the weights of the energy losses are lower than those of Section 6.10.1 so that the point of economical indifference (see Figure 6.16) is $w_x = 33.3\,€/\text{m}^2$.

It is worth noting that the maximum power P_R is compatible with the capability chart of a single circuit #c1 (with ampacity of case α_4) highlighting very important features for generation delivery continuity.

Table 6.8 Composition of overall cost [M€/km]

	#a1	2#c1
(I)	0.35	3.5
(E)	1.04	0.41
(OM)	0.16	0.32
(D)	0.007	0.037
(T)	$0.1 \cdot w_x$	$0.018 \cdot w_x$
(Σ) = (I) + (E) + (OM) + (D) + (T)	$1.56 + 0.1 \cdot w_x$	$4.27 + 0.018 \cdot w_x$

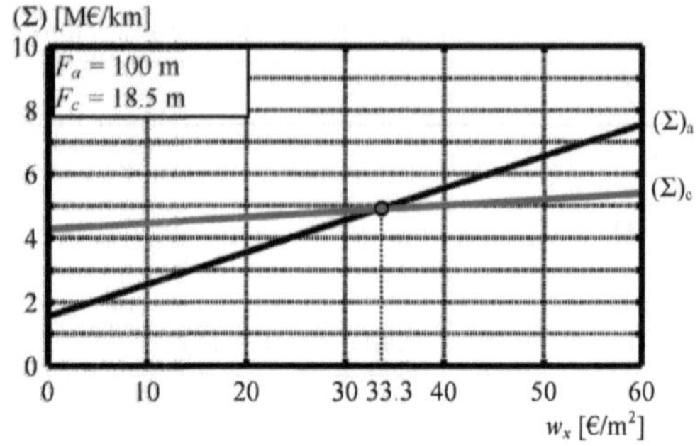

Figure 6.16 Overall costs per kilometre (with target of quality 3 μT)

6.10.3 Sensitivity to the Principal Parameters

Figure 6.17 shows (Σ) as a function of different exposure limits set by countries for a location with $w_x = 24.3\,€/\mathrm{m}^2$ which could occur in a rural area.

Once again, the UGC provides a rather favourable solution when restrictive limits of B are compelled (in Figure 6.17 for $0.2\,\mu\mathrm{T} \leq B \leq 3.0\,\mu\mathrm{T}$).

Figure 6.18 highlights the sensitivity of (Σ) to the ratio of the investment costs $(\mathrm{I})_c/(\mathrm{I})_a$ ranging between 8 and 10 and allows verifying how the variations of UGC investment cost (due to possible trader allowances, scale effect) play a key role in the comparative evaluations.

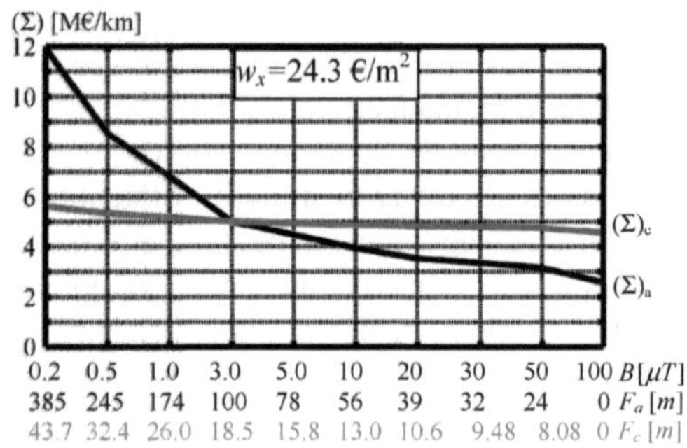

Figure 6.17 Cost sensitivity to B limit in territory with $w_x = 24.3\,€/\mathrm{m}^2$ and duration curve of Figure 6.14a

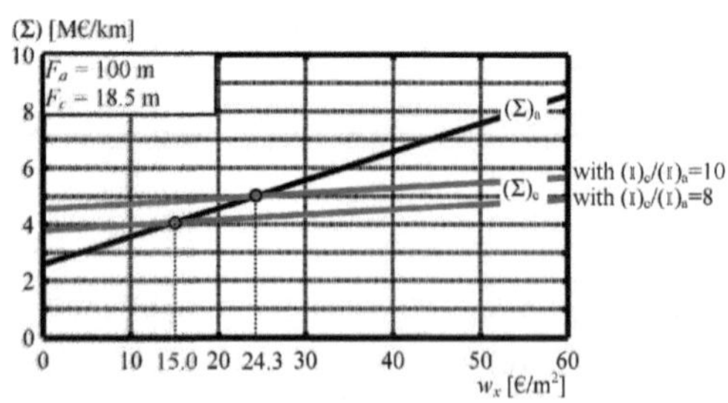

Figure 6.18 Cost sensitivity to $(\mathrm{I})_c/(\mathrm{I})_a = 8$–$10$; (with target of quality $= 3\,\mu\mathrm{T}$) and duration curve of Figure 6.14a

6.11 Case Study of Section 6.9 with Duration Curve of Figure 6.14a

The compensation power Q_{Mvar}, necessary on the whole for the double-circuit line 2#c1 (with $\xi_{\text{sh}} = 0.53$), can be estimated about

$$Q_{\text{Mvar}} = 0.53 \cdot (11 \cdot 2)_{[\text{Mvar/km}]} \cdot 25_{[\text{km}]} = 291.5 \quad [\text{Mvar}] .$$

Detailed evaluations of the whole system at both ends (where the cost of i) is the prevailing one) have brought to $\Delta_{\text{sh}} \cong 4.96$ [M€] corresponding to a kilometric investment cost $(\Delta\text{I})_{\text{Sh}} \cong \frac{\Delta_{\text{sh}}}{d} = 0.198$ [M€/km].

Table 6.9 shows the composition of overall costs where the weight of $(\Delta\text{I})_{\text{Sh}}$ is clearly detectable; Figure 6.19 highlights the point of economical indifference $w_x = 27.4\,€/\text{m}^2$ slightly greater than that of Figure 6.15.

It is worth remembering that for $d > 25$ km intermediate shunt compensations and greater ξ_{sh} are needed as explained in Chapter 3.

Table 6.9 Composition of overall cost [M€/km]

	#a1	2#c1
(I)	0.35	3.5
(E)	2.096	0.801
(OM)	0.16	0.32
(D)	0.007	0.037
$(\Delta\text{I})_{\text{Sh}}$	0	0.198
(T)	$0.1 \cdot w_x$	$0.018 \cdot w_x$
$(\Sigma) = (\text{I}) + (\text{E}) + (\text{OM}) + (\text{D}) + (\Delta\text{I})_{\text{Sh}} + (\text{T})$	$2.61 + 0.1 \cdot w_x$	$4.86 + 0.018 \cdot w_x$

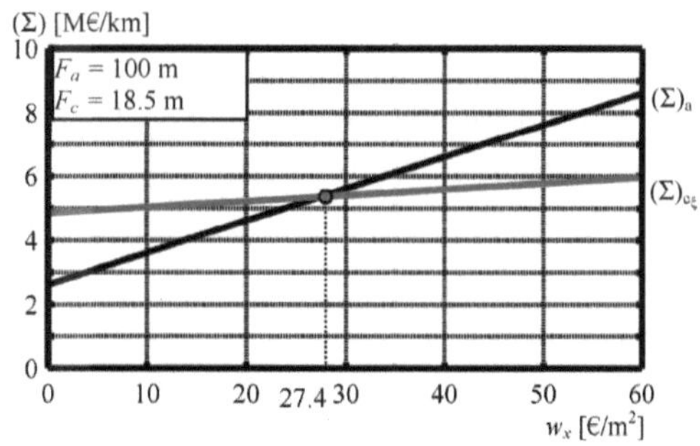

Figure 6.19 Overall costs per kilometre with compensated UGC ($d = 25$ km) and year duration curve of Figure 6.14a

6.12 Conclusion

In this chapter, the importance of an *overall cost evaluation* between UGC and OHL has been clearly highlighted.

Ensuring reliable and economic connections and respecting the environment is a crucial task, often requiring innovative solutions.

Whereas overhead lines have been the selected solution, extra high voltage cables have made a breakthrough with the introduction of XLPE insulation reducing dramatically the losses whilst maintaining an excellent performance level.

In spite of the low investment cost of overhead lines, UGCs have other important tangible benefits as well as some advantages which are less tangible.

This chapter makes a detailed technical and economic assessment of these two different technological solutions, considering the best *social* and economic solution in different conditions and shows that the evaluation must be carried out on a case by case basis.

The general procedure and approach have been reported by presenting a particular Italian situation as an example. The model however can be applied to any country with its own specific transmission standards, rules and/or laws.

In general, overhead lines and cables have been debated as competitors in a comparison without precise rules. It is apparent that where land has already been developed for residential use or where development potential is very high, underground cables are the preferred option having less environmental impact even if there are higher capital costs. Conversely, in areas of low value overhead line transmission having a greater environmental impact and lower capital costs remains the preferred option.

The authors are aware that a "comprehensive verdict" in the technical, economical, social and environmental fields is highly difficult and complex; nevertheless, they believe that the present schematic procedure can be deepened and enlarged to other aspects chiefly if it is performed during the planning phase when the different constructive, environmental, social and financial parameters can be accurately assessed.

More and more in the future the overall cost comparison between OHL and UGC will have to be evaluated: the authors hope to have contributed in constructing the engineering basis of this comparison.

References

1. Cigré Technical Brochure No. 110, Joint Working Group 21/22-01: Comparison of high voltage overhead lines and underground cables, December 1996
2. Directive 85/337/EEC: Assessment of the effects of certain public and private projects on the environment, Brussels, EC, 1985
3. R. Benato, D. Capra, R. Conti, M. Gatto, A. Lorenzoni, M. Marazzi, G. Paris, F. Sala: Methodologies to assess the interaction of network, environment and territory in planning transmission lines, *Cigré 2006, Paper C2-208*

4. G.V. Armani, R. Benato, C. Di Mario, A. Lorenzoni, M. Rebolini, F. Renaud: Application of multi-criteria methodology for planning an electric transmission line, *Cigré 2008, Paper C3-303*
5. Law 22 February 2001, n. 36: Framework Law on the protection against exposure to electric, magnetic, and electromagnetic fields, published on Official Gazette of the Italian Republic n. 55, March 7, 2001 (in Italian)
6. Decree of the President of the Council of Ministers, DPCM 8 July, 2003: Establishment of exposure limits, attention values, and quality goal sto protect the population against power frequency (50 Hz) electric and magnetic fields generated by power lines, published on the Official Gazette of the Italian Republic n. 200, August 29, 2003 (in Italian)
7. Decree 29 May 2008: Approval of the computation methodology for the determination of the right-of-ways of power lines (in Italian)
8. CEI 11-60 (Italian Standard): Carrying capacity at thermic limit of overhead lines exceeding 100 kV, 2002 (in Italian)
9. R. Benato, A. Paolucci: Voltage asymmetries in the electric network: connections of new power plants. *Proc. of General Meeting AEI*, Padova 3–5 October 2001, pp. 49–54 (in Italian)
10. L. Paris, M. Sforzini: Energy losses in the economic design of AC transmission systems, *World Power Conference*, Lausanne, 1964, Paper 92
11. C. Gary, M. Moreau: L'effet de couronne en tension alternative, Eyrolles, Paris, 1976
12. IEC 60287: Electric cables – Calculation of the current rating, (in 8 parts: 1.1, 1.2, 1.3, 2.1, 2.2, 3.1, 3.2, 3.3).
13. IEEE Std 442-1981 (reaffirmed 1996): IEEE GUIDE for Soil Thermal Resistivity Measurements
14. G.J. Anders: Rating of electric power cables in unfavourable thermal environment, IEEE Press, Wiley-Interscience, New Jersey 2005
15. R. Benato, D. Capra, R. Conti, A. Lorenzoni, E. Pinton, D. Smaniotto, R. Turri: A tool for the comparative analysis of the technical, economical, environmental and territorial features concerning transmission lines between the Italian grid and European and Mediterranean ones, Final Report CESI A4523730, January 2005 (in Italian)
16. R. Benato, M. Del Brenna, C. Di Mario, A. Lorenzoni, E. Zaccone: A new procedure to compare the social costs of EHV-HV overhead lines and underground XLPE cables. *Cigré 2006, Paper B1-301*
17. R. Benato, L. Fellin, A. Lorenzoni, A. Paolucci: Transmission systems in the territory. *AEI*, Vol. 88, No. 12, December 2001, pp. 28–35 (in Italian)
18. R. Benato, M. Del Brenna, C. Di Mario, A. Lorenzoni, E. Zaccone: Economical comparison between XLPE-cables and overhead lines, *Proc. of AEIT-Cigré*: The role of XLPE cables in the EHV electric energy transmission, Padova, 18 February 2005 (in Italian)
19. N. Wertheimer, E. Leeper: Electrical wiring configurations and childhood cancer. *Am. J. Epidemiol.* 109:273–284, 1979
20. S. Ueno: Biological effects of magnetic and electromagnetic fields. New York: Plenum Press; 1996
21. T.S. Tenforde: Biological interactions and potential health effects of extremely-low-frequency magnetic fields from power lines and other common sources, *Ann. Rev. Public Health* 13:173–196, 1992
22. ICNRP: Guidelines for limiting exposure to time-varying electric, magnetic and electromagnetic fields (up to 300 GHz), *Health Physics* Vol. 74, N. 4, April 1998, pp. 494–522
23. Swiss Federal Council: Ordinance relating to Protection from Non-Ionising Radiation (ONIR), Bern, 23 December 1999
24. Guide CEI 106-11: Guide for the determination of the respect widths for power lines and substations according to DPCM 8 July 2003 (Clause 6). Part 1: Overhead lines and cables, February 2006 (in Italian)

25. Technical Brochure Cigré No. 21: Electric and Magnetic Fields Produced by Transmission Systems, 1980
26. Technical Brochure Cigré No. 104: Magnetic Field in HV cable systems 1/Systems without ferromagnetic component, 1996
27. IEEE Magnetic Fields Task Force: Magnetic fields from electric power lines. Theory and comparison to measurements, *IEEE Trans. Power Delivery*, Vol. 3, pp. 2127–2136, Oct. 1988
28. Cigré Technical Brochure No. 373, Working Group C4-204: Mitigation techniques of power-frequency magnetic fields originated from electric power systems, February 2009
29. F. Kiessling, P. Nefzger, J.F. Nolasco, U. Kaintzyk: Overhead Power Lines. Planning, Design, Construction, Springer, 2003
30. G. Biondi, B.A. Cauzillo, L. D'Ajello, E. Guastadisegni, P. Nicolini, L. Paris, G. Piendibene: Quantitative evaluation of the visual impact of overhead lines, *Cigré 1996, Paper 22-210*
31. H.J. Koglin, R. Zewe, W.F. Bückner, F. Hirsch, K.H. Weck: Technical, economic and environmental aspects of compact overhead lines, *Cigré Symposium 33-91, Paper 700-07*, Leningrad 1991
32. R.C. Bishop, P.A. Champ, D.J. Mullarkey: Contingent valuation, The handbook of Environmental Economics, Basil Blackweel Ltd, Cambridge 1995
33. P.F. Colwell: Power Lines and land values, *Journal of real estate research* 5(1), pp. 117–127, 1990
34. S.W. Hamilton, G.M. Schwann: Do high voltage transmission lines affect property value?, *Land Economics* 71:4, 1995, pp. 436–444
35. R. Billinton: Canadian experience in the collection of transmission and distribution component unavailability data, *Proc. of 8th International Conference on Probabilistic Methods Applied to Power Systems*, IOWA State University, Ames, Iowa, September 12–16, pp. 268–273, 2004
36. Cigré Technical Brochure No. 385, Working Group B1-10: Up date of service experience of underground and submarine cable systems. May 2009
37. R. Benato, D. Napolitano: Reliability Assessment of EHV Gas Insulated Transmission Lines: effect of redundancies, *IEEE Trans. on Power Delivery*, Vol. 23, Issue 4, October 2008, pp. 2174–2181

Index

MIX
Papier aus verantwortungsvollen Quellen
Paper from responsible sources
FSC® C105338

If you have any concerns about our products,
you can contact us on
ProductSafety@springernature.com

In case Publisher is established outside the EU,
the EU authorized representative is:
Springer Nature Customer Service Center GmbH
Europaplatz 3, 69115 Heidelberg, Germany

Printed by Libri Plureos GmbH
in Hamburg, Germany